MEDIA AND COMMUNICATIONS - TECHNOLOGIES, POLICIES AND CHALLENGES

BROADCAST TELEVISION IN TRANSITION

MEDIA AND COMMUNICATIONS - TECHNOLOGIES, POLICIES AND CHALLENGES

Additional books in this series can be found on Nova's website under the Series tab.

Additional E-books in this series can be found on Nova's website under the E-books tab.

MEDIA AND COMMUNICATIONS - TECHNOLOGIES, POLICIES AND CHALLENGES

BROADCAST TELEVISION IN TRANSITION

DANIEL J. FETONNI
EDITOR

Nova Science Publishers, Inc.
New York

For permission to use material from this book please contact us:
Telephone 631-231-7269; Fax 631-231-8175
Web Site: http://www.novapublishers.com

Library of Congress Cataloging-in-Publication Data

Goldfarb, Charles B.
Broadcast television in transition / editor, Daniel J. Fetonni.
p. cm. -- (CRS report for Congress)
Includes index.
ISBN 978-1-62100-440-0 (softcover)
1. Television broadcasting--Economic aspects--United States. 2. Cable television--Economic aspects--United States. 3. Direct broadcast satellite television--Economic aspects--United States. I. Fetonni, Daniel J. II. Title.
HE8700.8.G635 2012
384.550973--dc23 2011034018

Published by Nova Science Publishers, Inc. † New York

CONTENTS

PREFACE

This book explores the transitions being made in broadcast television with a focus on the market forces affecting broadcast news networks and the Satellite Television Extension and Localism Act of 2010 (STELA).

Chapter 1- Market and technological changes are creating challenges to the long-standing business models employed by broadcast television networks and local television stations, but at the same time generating potential opportunities. The changes also may be affecting the three pillars of U.S. government media policy—localism, diversity of voices, and competition—and damping the effectiveness of existing regulations intended to foster them. These changes generally are strengthening the position of parties that own or control popular content in their negotiations with distributors of video programming. Broadcast networks and stations, alike, both own content and distribute programming, so they have been strengthened and weakened by these changes.

Chapter 2- The Satellite Television Extension and Localism Act of 2010 (STELA), P.L. 111-175, modifies the copyright and carriage rules for satellite and cable retransmission of broadcast television signals. The legislation was needed to reauthorize (through December 31, 2014) certain expiring provisions in the Copyright Act and the Communications Act and to update the language in those acts to reflect the transition from analog to digital transmission of broadcast signals, as well as to address certain public policy issues. Had the expiring provisions not been reauthorized, satellite operators would have lost access to a statutory compulsory copyright license and to statutory relief from retransmission consent requirements. This would have made it difficult, if not impossible, for them to retransmit certain distant broadcast signals to their subscribers, including signals providing otherwise unavailable broadcast network programming.

In: Broadcast Television in Transition
Editor: Daniel J. Fetonni
ISBN: 978-1-62100-440-0

Chapter 1

HOW CHANGES IN THE ECONOMICS OF BROADCAST TELEVISION ARE AFFECTING NEWS AND SPORTS PROGRAMMING AND THE POLICY GOALS OF LOCALISM, DIVERSITY OF VOICES, AND COMPETITION*

Charles B. Goldfarb

SUMMARY

Market and technological changes are creating challenges to the long-standing business models employed by broadcast television networks and local television stations, but at the same time generating potential opportunities. The changes also may be affecting the three pillars of U.S. government media policy—localism, diversity of voices, and competition—and damping the effectiveness of existing regulations intended to foster them. These changes generally are strengthening the position of parties that own or control popular content in their negotiations with distributors of video programming. Broadcast networks and stations, alike, both own content and distribute programming, so they have been strengthened and weakened by these changes.

---

* This is an edited, reformatted and augmented version of a Congressional Research Service publication, CRS Report for Congress R41458, from www.crs.gov, dated October 20, 2010.

The successful entry of hundreds of cable networks and the proliferation of social networking and video Internet websites have fragmented audiences and provided advertisers with alternative avenues for reaching consumers. This presents a significant challenge to broadcast networks and stations, which traditionally have relied on advertising for more than 90% of their revenues. As audiences have declined for both national and local news programming, networks and stations alike have reduced costs by sharing the fixed costs of newsgathering over multiple platforms and undertaking cooperative newsgathering with other outlets. Some broadcasters have sought to generate additional cost savings or revenue by combining with other newsgathering organizations, and urge modifications to the Federal Communications Commission's broadcast media ownership rules that restrict such combinations. Policymakers will have to weigh whether allowing such consolidation will, on net, benefit the public by improving the financial viability of newsgathering firms or harm the public by reducing diversity of voices and competition.

At the same time, competition has developed among the companies that deliver multiple channels of video programming to subscribers—cable operators, satellite operators, and some telephone companies. If a multichannel video distributor fails to obtain the retransmission rights to popular national and local broadcast television programs, it risks losing subscribers to a competitor that does offer the programming. As a result, broadcast networks and stations are able to demand higher payments from these multichannel video distributors for the retransmission rights. This has created a second revenue stream for broadcasters that is projected to reach $2.6 billion in 2016. It also occasionally results in subscribers losing access to broadcast programming when their video provider and the broadcaster reach an impasse in retransmission negotiations. A coalition of video distributors and consumer organizations has petitioned the FCC to modify its retransmission consent rules by adding dispute resolution mechanisms and mandatory interim carriage.

The amount of local broadcast news programming has been increasing despite declining audiences and does not appear to be threatened by stations' revenue declines. Many stations are broadcasting more news because local news programs can be cheaper to provide than purchased programming. In addition, local news provides a way for stations to develop strong brand identities as they compete for local advertising dollars.

The production and distribution of major sports programming is largely controlled by the sports entities that control the events. If it benefits them to distribute their programming through pay venues, such as cable networks that they own, rather than over the air broadcast, they will do so. This is likely to result in more events being televised, though many will only be available to subscribers to pay television services.

Overview

Market and technological changes are creating challenges to the long-standing business models employed by broadcast television networks and local television stations, but at the same time generating potential opportunities for those networks and stations. These changes generally are strengthening the position of parties that control popular program content in their negotiations with distributors of that programming. The changes also may be affecting the three pillars of U.S. government media policy—localism, diversity of voices, and competition—and damping the effectiveness of existing regulations intended to foster them.

As a result, various stakeholders have asked the Federal Communications Commission (FCC or Commission) to eliminate or modify some of its broadcast media ownership rules and retransmission consent rules. In particular:

- The FCC currently is collecting data and comments, and sponsoring research, as part of the statutorily mandated quadrennial review of its broadcast ownership rules.[1] Those rules, among other things, set limitations or prohibitions on the number of national broadcast networks under a single owner, the number of television stations that can be jointly owned in a local market, and the cross-ownership of a television station and a major daily newspaper in a local market.[2] In its comments in that proceeding, the National Association of Broadcasters (NAB) seeks elimination of the cross-ownership restrictions and relaxation of the restrictions on a single entity owning or controlling two television stations in a single local market, claiming these rules prevent efficient combinations required to support the cost of providing local news and emergency journalism.[3] Other commenters claim that such consolidation would harm, rather than foster, diversity of voices, competition, and localism. For example, the National Association of Black Owned Broadcasters claims that consolidation places smaller firms with few stations at a competitive disadvantage, and that " Any further relaxation of the Commission's multiple ownership rules would exacerbate the already dismal lack of minority ownership in the broadcast industry."[4]

- The FCC is seeking comment on a petition for rulemaking, submitted by a coalition of multichannel video distributors[5] (cable operators, satellite operators, and telephone companies) and consumer organizations, requesting that the Commission amend its retransmission consent rules[6] to include dispute resolution mechanisms and mandatory interim carriage while disputes are being resolved.[7] The NAB and the associations of the local station affiliates of the four major broadcast network oppose the petition, claiming that the market is working properly and that FCC intervention is unwarranted and would be harmful.[8]

Since the FCC has addressed such issues in response to statutory instructions intended to foster diversity, competition, and localism, Congress may choose to provide statutory or informal guidance to the FCC as it moves forward with the proceedings.

The Market Changes

Broadcast networks and broadcast stations control program content that is highly valued by multichannel video distributors and also distribute programming in competition with many other media outlets. Thus, they are both strengthened and weakened by recent market changes.

Most notably, the successful entry of hundreds of cable networks and the proliferation of social networking and video Internet websites have fragmented audiences and provided advertisers with alternative avenues for reaching consumers. A recent UBS Investment Research report[9] claims that "every year, global advertising is losing 200 points [2 percentage points] of growth to other marketing usage or is eliminated, thanks to technology improvements" and that "structural changes [affecting broadcast television advertising] remain strong and are even accelerating." This presents a significant challenge to broadcast networks and stations, which traditionally have relied upon advertising for more than 90% of their revenues.[10] (In contrast, in 2008, advertising revenues represented only 42.8% of the revenues of advertising-supported cable networks[11] and that proportion likely has fallen since then.)

Several research companies project television advertising revenues, each employing its own proprietary model and assumptions.[12] Although the various forecasters project different levels of advertising revenues, they generally identify the same trends. As an example, according to a recent forecast by Magna Global, a consultant to advertisers:[13]

- the advertising revenues of *English language national broadcast networks* peaked at $14.7 billion in 2005, fell slightly in 2006 and 2007 and then fell significantly in 2008-2009, have begun to recover in 2010 but are not likely to reach the previous peak until about 2015.
- the advertising revenues of *Spanish language national broadcast networks* experienced a slight decline from their 2007 peak but are growing rapidly and may be 50% higher in 2015 (exceeding $1.5 billion) than they are today.
- the Olympics, which have provided a revenue shot in the arm in alternate years to the network with the broadcasting rights and its affiliate stations, generating as much as $785 million in incremental advertising revenues in 2000, will generate substantial, but lesser amounts in the future as more of the programming is carried on cable networks and the Internet.[14]
- the non-political advertising revenues of *local broadcast television stations* peaked in 2007 at about $17.8 billion, fell substantially in 2008 and 2009, and are now experiencing increases, but are not expected to reach the 2007 high by 2015.[15]
- the political advertising revenues of local broadcast television stations, which are much higher in even numbered years than in odd numbered years, and greatest for the quadrennial presidential election, will continue to grow substantially, so that the sum of non-political and political advertising revenues are likely to equal or surpass those of the earlier peak. But political advertising revenues are not spread equally across all broadcast stations; they are focused on stations in geographic areas where there are contested elections.

This forecast implies that during the next five years, total broadcast television advertising revenues will at best recover to their previous peak.

Other market changes, however, have led to a brighter outlook for broadcasters' non-advertising revenues. For example, competition has developed among the companies that deliver multiple channels of video programming to subscribers—cable operators, satellite operators, and some telephone companies. If a multichannel video distributor fails to obtain the retransmission rights to national and local broadcast television programming, which many households consider "must have" programming, it risks losing some of its subscribers to a competitor that does offer the programming. As a result, broadcast networks and stations are able to demand higher cash payments from these multichannel video distributors for the retransmission rights. This has created a second revenue stream for broadcasters that is projected by SNL Kagan, a media research firm, to grow from $762 million in 2009 to $1.09 billion in 2010, $1.36 billion in 2011, and more than $2.6 billion in 2016.[16] Note, however, that the projected increases in retransmission consent revenues, which will be shared by broadcast networks and local stations, may be less than the fall in network and local station advertising revenues from their earlier peaks.

Technological innovations are creating the potential for other new broadcast revenue sources. For example, in June 2010 twelve large broadcast groups created a joint venture, Mobile Content Ventures, to develop broadcast-based mobile digital television technology and bring it to market.[17] In September 2010, a second consortium of broadcasters owning 346 television stations announced the creation of the Mobile500 Alliance in support of mobile digital television technology.[18] This technology, an alternative to wireless broadband video, would allow a station to broadcast encrypted video signals that could be received by anyone in its geographic service area with the appropriate receiver; the viewer could move anywhere within the service area and continue to receive the signal. With broadcast mobile video, the bandwidth (and, hence, spectrum) requirements increase as the number of different video signals offered increases, but do not increase as the number of users increases. In contrast, with wireless broadband video distributed to cell phones, the more users in a location, the more bandwidth and spectrum is needed. Since the broadcast technology allows many consumers to share the same spectrum, broadcasters are unlikely to have to impose the sort of usage charges that AT&T recently imposed on its customers.[19] The underlying broadcast technology may not allow broadcasters to offer as wide an array of programming as wireless broadband video providers, but the broadcasters already own or control much of the "must-have" programming sought by consumers.

Also, with the digital transition, broadcasters are able to broadcast multiple video programming streams, rather than a single stream, and this "multicasting" may generate additional advertising and retransmission consent revenues for broadcasters. Although to date revenues are small, some analysts are bullish on multicasting's potential impact on broadcast station cash flow.[20]

None of these potential new revenue streams has been proven yet in the marketplace, and other technological innovations may not benefit broadcasters. For example, while it is possible that the nascent 3-D technology will lead to popular programming, it is more likely that such programming will be distributed via a pay-television model than an over-the-air broadcasting model.[21]

News, Sports, and Broadcaster Profitability

This macro picture of the broadcasting industry, with some networks and local stations facing declining advertising revenues without compensating new sources of revenues, has led to concern about the future of broadcast television news and sports programming.

More Americans rely upon television news than any other news medium[22] and, despite the success of cable news networks, the national news broadcasts of ABC, CBS, and NBC continue to attract more prime time viewers than the cable networks.[23] But most network news programs continue to lose audience share,[24] and the average viewer age for the networks' evening newscasts has risen to 63.1 years.[25] Not only does a smaller audience mean that broadcasters can charge less for ads, but as advertisers tend to prefer younger viewers, an aging audience generates less advertising revenue. As a result, in the past three years all three national broadcast networks that have national news divisions have made cuts in those divisions and taken other steps that could affect the quality or scope of their news coverage.[26]

Although viewership of local television news programming continues to decline,[27] Americans still rely more on broadcast television than any other media source for local news and public affairs information.[28] Those stations that produce original news programming are broadcasting more such programming than ever,[29] but there has been a slight decline in the number of stations that produce news programming and some stations have cut their news staffs.[30] Some of the additional hours of news consist of repeats of earlier broadcasts rather than fresh programming.

Increasingly, television stations that create news programming sell some of their programming to other media outlets.[31]

Major professional and intercollegiate sports leagues command very high payments for the rights to televise their sports events. Cable networks have been able to successfully bid for the programming rights to major events, either in conjunction with or in direct competition with broadcast networks. Most recently, the National Collegiate Athletic Association (NCAA) signed a $10.8 billion, 14-year agreement with CBS and Time Warner's Turner Cable Networks under which Turner will get the rights to first- and second-round NCAA basketball tournament (March Madness) games in 2011-2015. Coverage of the semifinals will be split between Turner and CBS, with the national championship game alternating each year between the two networks.[32] At the same time, major sports leagues and conferences increasingly are retaining a portion of their sporting events for cable networks in which they have an equity stake, rather than making them available to outlets in which they do not have an equity stake. The National Football League (NFL), the Big Ten Intercollegiate Conference, and the New York Yankees have created cable networks of their own and distribute some of their games exclusively or semi-exclusively over those networks. These changes have increased the total amount of sports programming available to U.S. households, but an increasing proportion of popular teams' or leagues' schedules is televised only on cable networks and therefore available only to subscribers.

Some broadcasters claim that relaxing broadcast media ownership rules would assure the continued availability of free, over-the-air news and sports programming. They also claim that their ability to offer such programming would be harmed if changes to the retransmission consent rules constrain their ability to negotiate compensation from multichannel video distributors. A report prepared for the National Association of Broadcasters, based on an April 2010 survey of 53 television stations that originate local news programming,[33] concludes:

> As the Commission examines the Future of the Media and considers ways to bolster the provision of local news, it should adopt policies that allow (and to rescind, or at least not adopt, policies that hinder) local broadcasters to (1) pursue opportunities for non-advertising revenue, such as that derived from retransmission consent, and (2) benefit from economies of scale and allocate their news resources in the most efficient ways, such as through modifications to the Commission's structural ownership rules. Such policies will support the Commission's vital focus on "localism" by providing a solid financial foundation for the production of local news.

Not all market data, however, suggest that broadcast provision of news and sports programming is at risk. News and sports programming are major sources of station revenues and continue to be profitable for most broadcast networks and stations.[34] Moreover, it often is easier for broadcast networks or stations to control the costs of news programming they produce than the costs of scripted or syndicated programming they must purchase. As will be explained below, broadcasters at both the network and local station level have undertaken cost-cutting measures and are employing business models that better exploit the economies of scope associated with newsgathering and program production. These changes may require Congress and the FCC to review how best to foster the goals of localism, diversity, and competition.

Audience fragmentation and the substantial upfront fixed costs associated with newsgathering and the rights to major sports events are not going away. Increasingly, to recoup these upfront costs, programming will have to be distributed over multiple platforms, and media outlets may have to share facilities and staff to gain economies of scope and scale. But some broadcasters claim that cooperative activity and sharing does not provide the cost savings or generate the additional revenues attainable if they could combine with other newsgathering organizations. Traditionally, there have been concerns that such consolidation could reduce diversity of voices, harm competition, and discourage programming of local interest. In light of the underlying market changes, however, Congress and the FCC may decide to reexamine whether and when the benefits of consolidation of newsgathering capabilities outweigh the harms and whether it would be possible to construct either bright-line rules or criteria for case-by-case evaluations that could limit the harms while retaining the benefits.

Market Forces Affecting Broadcast News Networks

Two studies of national broadcast television news networks prepared 11 years apart—*The State of the News Media 2010* study prepared by the Pew Foundation Project for Excellence in Journalism and a Nieman Foundation Special Issue report prepared in 1999[35]—present very similar portraits of the market forces at play and identify the same market trends and policy implications. This suggests that many of the market forces that broadcast news networks are contending with today represent long-term structural trends that may be accentuated by, but are not the result of, cyclical economic factors. It also suggests that while these trends have challenged old business models, the networks have been able to adapt and to survive in the marketplace.

Three of the four major national broadcast networks—ABC, CBS, and NBC, but not FOX—have national news divisions that produce for their local affiliates half-hour evening newscasts, morning news shows (that range in length from two to four hours and tend to provide softer news than the evening newscasts), Sunday morning public affairs discussion shows, weekly or periodic news magazine shows (such as "60 Minutes" and "20-20") that mix investigative news with softer news and entertainment, and unscheduled breaking news reports.[36] This mix of programming has not changed significantly in the past decade. Although the networks continue to lose audience to cable and Internet providers of news and information, they still serve a much larger audience than the cable news networks or Internet news services.[37]

Interestingly, in the late 1990s the broadcast television networks were losing money on their entertainment and sports programming, for which costs were growing faster than revenues, but were able to earn profits on their news programming, in part by controlling costs.[38] NBC News, the most watched and the most profitable national broadcast news division, a decade ago created a multiplatform organization that shares the high fixed costs of newsgathering across broadcast, cable, and Internet platforms. NBC News was quite profitable then and remains so.[39] The other two national news divisions, ABC and CBS, do not have cable news networks with which to share costs and have not developed as effective an Internet presence. They earned relatively small profits in 1999 and today they are at best marginally profitable and, in the case of CBS News, may be losing money.[40]

The *Nieman Report* and *Pew Study*, supplemented by current market data, provide a dynamic portrait of the broadcast network news market.

- The average nightly audience for the three major broadcast networks' evening newscasts has fallen from more than 50 million viewers in 1980 to 30.4 million in 1998 to 22.3 million in 2009, a decrease of about 1 million viewers per year (though the annual decline has been less than that in recent years).[41] The audiences nonetheless remain far larger than the prime time audiences for cable news networks.[42] The audience for broadcast network morning news shows also has fallen for the past five years.[43] The prime time news magazine programs continue to attract large audiences, with each of the three networks having at least one successful program.

- The audiences for the evening newscasts of the two Spanish language broadcast networks—Univision and Telemundo (owned by NBC)—are almost as large as the prime time audiences for the major cable news networks (FOX News, CNN, MSNBC, HLN, and CNBC).
- The declining and aging audience for evening and morning newscasts has placed downward pressure on advertising revenues. This is countered somewhat by the relative affluence of that audience and by the fact that, despite the decline, the audience for the evening newscast at each network continues to be significantly larger than the audience for any other broadcast or cable programming presented at that time of day. The network newscast audiences therefore continue to command a premium from advertisers. There has been a shift toward sponsorships of particular segments of network news programs—or even entire newscasts—by a single advertiser. Still, advertising revenues generated by evening and morning network news programming have been falling.
- Since the 1990s,[44] there have been periodic layoffs in all three network news divisions. Some domestic and foreign bureaus were closed, with the networks relying more heavily on outside sources they could pay as needed rather than on full-time staff and offices. The broadcast networks have expanded their use of "pooling"—sharing footage from a single camera with other networks, including cable networks. They have reduced the amount of "breaking news" coverage, increasingly leaving that to 24-hour cable news networks. These cost-cutting measures continue. In 2010,[45] ABC announced it would reduce its news division by 300 to 400 employees, out of 1,500. CBS News reduced its staff in 2010 by 100 people, or 7%. NBC News reduced its workforce by more than 10% in 2009. Although the number of domestic bureaus has stabilized, the three networks continue to trim permanent overseas bureaus and rely more on one-person operations, often sending resources to overseas locations only as events occur.
- The broadcast news networks continue to employ a business strategy developed in the 1970s, using news teams to create hybrid news and entertainment programming shown during prime time. These "newsmagazine" programs tend to be less expensive to produce than scripted entertainment programming for two reasons: many of the

underlying costs can be shared with the hard news programming and, even when successful, the correspondents and writers tend not to be able to demand the huge increases in salary and other compensation that the stars and writers of successful entertainment programming can demand. As a result, even if newsmagazines generally attract smaller audiences than pure entertainment programming, they can be more profitable. The correspondents and news teams have credibility from their involvement with hard news programming that their counterparts on competing cable networks may lack. Newsmagazines continue to provide the financial backbone of network news organizations.

Even with serious cost containment measures in place, the fixed costs associated with maintaining a news network are substantial. Since the late 1990s, the networks have responded by seeking economies of scope by using those news assets to produce revenue-generating programming for multiple platforms. The 1999 *Nieman Report* explained how NBC was already successfully following this business model:

> Here is why this model works so well. ABC and CBS generate nearly all of their income from broadcasting. NBC takes in additional revenues from cable (MSNBC, CNBC), NBC's Internet sites, and NBC News channels distributed outside of the United States.... Therefore NBC is better able to shoulder the overall costs involved with newsgathering not just for the "Nightly News" and "Today," but for its 24-hour cable networks and Internet sites as well....By programming more than 6,000 hours of news a year across its broadcast and cable platforms, the average cost per hour of news at NBC has fallen from about $250,000 to $50,000 during the past five years.[46]

This essence of this quote is echoed in the 2010 *Pew Study*, which finds:

> NBC News is a profit center for NBC Universal, but its earning power now comes from its multiplatform structure. On paper, the news division derives more revenue from cable news channels than from the broadcast. About two-thirds of news revenues come from MSNBC and CNBC. ABC News and CBS News, by contrast, have no such multiplatform systems....
>
> Sooner rather than later, other platforms like the Web or video streaming will need to become a more integral part of the networks'

distribution and revenue strategy, a strategy NBC has already largely developed and reaped benefits from.[47]

Market forces thus appear to be pushing ABC and CBS toward partnerships with other news organizations. The *Pew Study* cites speculation in 2010 that ABC or CBS were looking to partner with cable news networks, with CBS News and CNN the likeliest partners since CNN correspondents already appear regularly on the CBS program "60 Minutes."[48] In 2009, ABC reporters began appearing on the Bloomberg business cable channel, the two organizations have jointly hired employees, and Bloomberg obtains content from ABC's affiliate news service. It also is possible that the ABC or CBS news divisions could seek to form partnerships with non-television newsgathering organizations, such as major newspapers, to share the high fixed costs of newsgathering. Such consolidation might prove the best way to ensure the continued existence of three strong newsgathering organizations supporting broadcasting networks, though at the potential expense of diversity of news sources and competition.

Formal Ties between National Newsgathering and Programming Organizations: Issues for Congress

A formal tie between two previously independent major newsgathering organizations would trigger review by either the Antitrust Division of the Department of Justice or the Federal Trade Commission to determine whether the combination would lessen competition. This review would only scrutinize economic effects—for example, if the combination would allow the new entity to raise the price broadcast stations and other outlets pay for news programming. It would not consider the impact on the diversity of voices or localism.

If the proposed merger or formal tie would result in the transfer of ownership or control of a spectrum license—broadcast networks and other major newsgathering organizations are likely to hold a variety of licenses for spectrum needed to gather and distribute information and programming—the FCC would have to review the proposed transfer to determine whether it is in the public interest. That review would take into account such public interest goals as diversity of voices and localism, as well as competition.

Although the merger of two major newsgathering organizations primarily would affect diversity and competition at the national level, such a combination also would be subject to existing FCC broadcast media ownership rules for local markets if the merging companies owned local outlets, as all the major broadcast networks, and some other major newsgathering organizations, do.

The national cable news networks do not have a local presence and therefore a merger with a national cable news network would not be subject to the FCC's broadcast media ownership rules. There are no specific statutory or FCC rules that set bright line tests or even generic criteria for evaluating the public interest implications of a proposed combination of a broadcast news network and a cable news network.[49] Given that both the NBC and FOX corporate organizations already own broadcast newsgathering capabilities alongside cable news networks without being subject to unique regulatory conditions or requirements, it might be difficult for the FCC to set conditions on an ABC or CBS merger with a cable news network.

It is possible that a broadcast news network seeking to share the fixed costs of newsgathering and to exploit economies of scope would find a good potential partner among the major newspaper companies, which face similar, if generally more threatening, challenges.[50] Most large newspaper companies already own television stations and websites that use video, so they already have some experience turning their newsgathering capabilities into video news programming. They also have a presence in many local markets, though perhaps not in the local markets where ABC or CBS own their own local affiliate stations.

If ABC News or CBS News were to propose a merger or other formal tie with a major newspaper company, then, in addition to the normal antitrust review and the FCC transfer of license review, the proposed combination would be subject to the current local market broadcast-newspaper cross-ownership rule.[51] The rule includes a rebuttable presumption that the merger in one of the 20 largest local markets of a major daily newspaper and one of the four highest-rated television stations would be deemed inconsistent with the public interest. ABC and CBS own and operate local stations in many large cities. In all of those cities, the ABC and CBS stations are among the four highest-rated in the market. It therefore might be necessary for the combined entity to divest itself of a television station or a newspaper to consummate the merger. In addition, most major newspaper companies also own broadcast television stations, which might compete in a local market with an ABC or CBS owned-and-operated station. In such a situation, the companies might be unable to consummate a proposed merger without divesting a local television station.

The local market broadcast ownership rules thus could make it more difficult for the ABC or CBS news networks to consummate a merger or formal tie with a major newspaper company than with a cable news network.

The economic impact of Internet competition and audience and readership fragmentation is affecting all newsgathering organizations, though some are more challenged than others. According to the *Pew* Study, the median prime-time viewership of the three major cable news networks (Fox, CNN, and MSNBC) increased from approximately 1.3 million households in 1998 to about 3.9 million in 2009, but in 2009 CNN's median prime-time audience fell 15% from 2008, while the Fox audience grew by 19% and the MSNBC audience grew by 3%.[52] The cumulative audience (the number of unique viewers who watch a channel for at least six minutes over the course of a month) of the three cable news networks fell slightly between 2002 and 2009, but remains between 50 million and 68 million for each of the networks.[53] Cable news network revenues and profits continue to be robust. The *Pew Study* reports that SNL Kagan estimated that aggregate revenues for Fox News, CNN, HLN, and MSNBC increased from $2.62 billion in 2008 to $2.76 billion in 2009, and profits increased from $1.07 billion in 2008 to $1.16 billion in 2009.[54] Although the combined revenues of CNN and HLN fell slightly in 2009, their profits increased slightly.[55]

In contrast, newspaper readership, revenues, and profits are falling. The number of daily newspapers in the United States fell from 1,600 in 1990 to 1,400 in 2008, [56] and continues to fall. Paid newspaper circulation in the U.S. in September 2009 was 43,500,000 daily and 46,500,000 Sundays, a decline of 31.5% and 27%, respectively, from peak totals in the last 25 years.[57] That decline has been greatest recently. In the six-month period ending September 30, 2009, compared to the same period a year earlier, newspaper circulation fell 10.6% daily and 7.1% Sundays. Newspaper advertising revenues reportedly fell by a total of 23% in 2007 and 2008 and by 26% in 2009.[58] The *Pew Study* estimates that, despite substantial cost cutting, average newspaper operating margins fell from the high teens in 2007 to the low teens in 2008, and then to around 8% in 2009.[59]

It appears that the newsgathering activities of newspaper companies are more challenged than those of cable news networks. But the current FCC broadcast media ownership rules, which focus on safeguarding localism, could unintentionally encourage broadcast network consolidation with newsgathering organizations such as cable news networks rather than with newspaper companies. This is consistent with longstanding U.S. media policy to foster localism and local programming, but could lead to a reduction of national and international news coverage.

Market Forces Affecting Local Broadcast News

Today, just under half the non-PBS broadcast television stations in the United States produce and air original local news programming, an additional 14% broadcast local news programming produced by another station,[60] and somewhat less than 40% do not broadcast local news at all.[61] This represents a small decline from 2009 in the number of stations originating news programming, but a slightly larger increase in the number of stations broadcasting news programming produced by others.

The Amount of Local Broadcast News Programming Is Increasing as Its Audience Decreases

Those stations that produce their own local news programming increased the amount of original news programming aired from an average of 4.7 hours per station per weekday in 2008 to an average of 5.0 hours in 2009, continuing an upward trend that began in the early 2000s.[62] Although most stations reported no changes in the amount of newscasts in 2009, 28.6% reported adding a newscast and only 13.7% reported cutting a newscast;[63] almost all of the cuts were in weekend programming. Some of the additional programming represents the rebroadcast of programming offered in an earlier program or the use of national or international news programming provided by the national network to which the station is affiliated.

Interestingly, the increase in local news programming comes when audiences for local early morning, early evening, and late news programs are falling, and when local television advertising revenues and local television station profits also are falling. The *Pew Study*[64] analyzed Nielsen Media Research data for 2009 and found:

- ratings, share, and viewership declined for the local news programming of the affiliates of all four major broadcast networks—which produce most of the local television news in the United

States—in all months and in all news timeslots studied. The viewership declines for the early evening and late news were steeper than in 2008.

- even Fox affiliates, which employ the strategy of broadcasting local news at off-hours (7 a.m. and 10 p.m.), experienced declining news viewership.
- local affiliate news in nontraditional timeslots, such as noon and 4 p.m., also experienced declining audiences.
- only a few independent stations had large enough audiences for Nielsen to track. These collectively attracted 9% of the news audience when their news programming competed in the news time slots used by ABC, CBS, and NBC affiliates. When they competed in the time slot used by FOX affiliates, they typically attracted at most 10% of the news audience. Only at noon did they capture almost half the news audience.

Broadcast Station Profits Fell Most Rapidly in the Late 1990s

The broadcasting industry has voiced concern that the decline in station advertising revenues without compensating new revenue sources could reduce profitability and impair stations' ability to offer news programming. It is difficult to construct a reliable database of the profits of individual television stations because most are owned by companies that own multiple stations or other interests; the reported profitability of any individual station thus depends heavily on how the parent company chooses to allocate non-station-specific costs and revenues to the station. Nonetheless, companies' reports on individual stations may provide useful information on trends in station profitability, unless there is reason to believe that station owners as a group have an incentive to over-assign or under-assign non-station-specific costs and revenues to the station.

Since the 1980s, the NAB has collected data on local station revenues, expenses, and profits, which it compiles each year in an annual *Television Financial Report.*[65] It appears that the dollar level of profits enjoyed by local broadcast television stations in the mid-1990s eroded sharply in the late 1990s—perhaps as a result of competition from cable channels.[66]

Table 1 reproduces a table included in the submission made by the NAB in the FCC's current media ownership proceeding,[67] showing the distribution of station pre-tax profits for 1998-2008. (The NAB has not yet released data for 2009.) The data show that profits (1) are highly cyclical, falling substantially in economic downturns, when advertising falls, and (2) generally increase in even-numbered years, when broadcasters benefit from high levels of political advertising and broadcasts of the Olympic Games. In each year, average profits are significantly higher than median (50th percentile) profits, and in many years are higher than the profits of the station in the 75th percentile, indicating that a relatively small number of highly profitable stations bring up the average.

Table 1. Broadcast Television Station Pre-Tax Profits All Markets, All Stations

Year	Average	25th Percentile	50th Percentile	75th Percentile
1998	$6,145,583	($220,970)	$1,575,778	$5,944,967
1999	$4,361,828	($659,146)	$916,554	$4,323,452
2000	$4,537,894	($584,884)	$1,113,634	$4,596,413
2001	$2,171,188	($1,445,544)	$67,067	$2,575,895
2002	$3,858,644	($451,601)	$911,827	$4,188,476
2003	$4,073,056	($458,512)	$464,019	$3,344,000
2004	$4,442,379	($158,079)	$1,128,782	$4,686,237
2005	$3,512,208	($512,639)	$670,946	$3,426,952
2006	$4,210,359	($305,161)	$1,120,443	$4,154,310
2007	$3,320,667	($454,837)	$520,164	$3,446,126
2008	$2,686,481	($750,149)	$630,300	$3,178,780

Source: Mark J. Prak, David Kushner, and Eric M. David, "The Economics Realities of Local Television News— 2010: A Report for the National Association of Broadcasters", *In the Matter of 2010 Quadrennial Regulatory Review—Review of the Commission's Broadcast Ownership Rules and Other Rules Adopted Pursuant to Section 202 of the Telecommunications Act of 1996*, Comments of the National Association of Broadcasters, Attachment C, p. 2, "Television Station Financial Data 1998-2008: Pre-Tax Profit Television Station National Averages," based on NAB Television Financial Surveys 1999-2009.

Average pre-tax profits have remained well below the high level posted in 1998. It is possible that some of that decline is attributable to high debt loads taken on by ownership groups that financed acquisitions of stations through highly leveraged borrowing.[68] Such leveraging increases the risk of financial loss if revenues do not grow as forecast, but increases the return on equity if revenues meet expectations. It is possible that a highly leveraged firm could experience a financial loss despite enjoying positive operating cash flow, because the latter takes into account operating expenses but not debt service.

The NAB's *Television Financial Reports* present data indicating that the great majority of stations enjoy positive operating cash flows. The 85 tables in the 2009 *Report* suggest, however, that there are three general categories for which a significant minority of stations suffered negative operating cash flow in 2008: major network affiliates in small markets, Spanish language stations in large markets, and independent stations.[69] While losses are sometimes attributable to poor management, there also may be other explanations. Some major network affiliates in small markets are owned by station groups that may be allocating non-station-specific costs to those stations. Alternatively, it may be that small markets have too little advertising to support all the stations in the market. Similarly, Spanish-language stations that are not affiliated with a major Spanish-language network may have difficulty attracting sufficient advertising revenues, even though they are located in a large market, because they attract a niche audience. The same is true for independent stations, which do not have access to network programming.

Even though most stations are enjoying positive operating cash flows, the high debt loads of some stations are imposing large interest expenses that may be causing pre-tax losses.[70] These debt burdens do not appear to be threatening the viability of the stations, which generally continue to enjoy positive operating cash flows. In the more serious cases they are forcing the financial restructuring of the station groups, with lenders taking equity positions while the stations continue to operate largely as before. Nonetheless, financial distress can affect station investment and programming decisions.

It would be useful to have a better understanding of the underlying financial position of those stations that are experiencing pre-tax losses. If most of those stations have positive operating cash flows, are affiliates of major national networks whose retransmission consent revenues are likely to grow,[71] and are successfully offering branded local news programming to their communities, but are burdened by interest expenses due to highly leveraged acquisitions, then there may be little reason for concern that losses will lead to reductions in local news programming. On the other hand, if most of the stations with pre-tax losses originate local news programming but have not successfully created strong brand identities, do not have major network programming to bolster their retransmission consent negotiations, and are suffering operating losses due to structural market changes, then it is more likely that local news programming will be cut back. It would be useful if the FCC could collect the data necessary to analyze these relationships as part of the current quadrennial review of its media ownership rules.

Falling profits and the expectation that revenues may be relatively slow to recover have reduced selling prices of broadcast television stations. SNL Kagan estimates that the "stick price" for a television station—roughly, the price per household that can be reached with a signal of reasonable quality for a station without measurable market share or cash flow, thus simply representing the value of the broadcast license and equipment[72]—peaked at $32.52 in 2001 and then fell irregularly to a low of $12.68 in 2009.[73] As other avenues became available for advertisers to reach consumers, the broadcast licenses generated less cash and became less valuable.[74] It is likely that most broadcast television stations will continue to generate revenues that exceed their operating costs, but the value that the broadcast license commanded when the station was a cash cow may continue to fall, or at least not rebound to earlier peaks. Owners that acquired licenses near the peak price may have difficulty covering those acquisition costs.

Financial Factors Specific to News Programming

While a station's decisions on the amount and type of local news programming may be affected by its overall financial status, those decisions are more likely to be affected by the direct financial impact of the news programming itself. Although broadcast stations are not required to report financial information specific to their news programming, there is some information available about the revenues, costs, and profits attributable to such programming.

For those stations that originate their own local news programming, that programming generates a significant portion of station revenues. Although most local station news directors responding to the *RTDNA/Hofstra Survey* did not know what portion of their station's revenues came from news programming, those that did know reported, on average, that local news generated 44.7% of station revenues—about the same level as reported throughout the past decade.[75] The median reported figure was 45% and the average and mean did not vary much by market size, never falling below 39.7% or rising above 50%. Since the amount of news programming has increased over time, however, this suggests that news programming is generating a smaller share of station revenues per hour of programming. The *NAB Survey* corroborates the importance of local news for station revenues. It found that "although local news programming accounts, on average, for only

16% of the broadcast day, 39% of a station's revenues, on average, is derived from advertising associated with the broadcast of local news."[76]

News programming continues to be profitable for most stations. The annual *RTDNA/Hofstra Survey* asks news directors about the profitability of their news programming. As shown in Table 2, far more news directors characterize their news programming as profitable than unprofitable or just breaking even. While a smaller percentage of respondents indicated in 2010 that their news programming was profitable than in most previous years, at the same time a smaller percentage of respondents indicated that their news programming lost money than in most previous years (reflecting, in part, that more news directors reported they did not know the profitability).[77] The percentage of stations reporting that their news programming is profitable did not vary much by market size—ranging from a low of 42.9% for stations in the 25 largest markets to a high of 51.7% for stations in markets 26-50. In the smallest markets (151-210), 45.8% of stations reported their news programming was profitable, 12.5% reported breaking even, 8.3% reported losses, and 33.3% did not know. Even for stations for whom news is not profitable, however, it may be beneficial to produce news programming if that programming creates a strong brand identity or if a news audience provides a stronger lead-in to the entertainment programming that follows.

Table 2. The Profitability of Local Station News Programming Percentage of Stations

	2010	2009	2008	2007	2006	2005	2004	2003	2002	2001	2000
Showing Profit	47.8	52.7	55.4	56.2	57.4	44.5	58.4	55.3	54.9	56	58
Breaking Even	14.6	12.6	11.5	11.5	8.1	24.2	10.4	13.6	11.6	13	11
Showing Loss	8.3	14.5	10.5	6.4	10.0	12.1	9.2	9.2	11.2	10	11
Don't Know	29.2	21.3	22.6	26.0	24.4	19.2	22.0	21.9	22.3	21	20

Source: *2010 RTDNA/Hofstra Staffing & Profitability Survey*, Section on TV News Staffing and Profitability, at p. 3.

The stations that spend most on local news production tend to be in the larger markets, but station spending on news programming does not appear to be correlated with market size and a higher proportion of stations in large markets do not originate local news. In its 2010 *Television Financial Report*, the NAB provides detailed information on the television news expenses (defined as "the salaries and wages of those engaged in the production and presentation of news and other expenses incurred by the news department") incurred by stations. Table 3 presents summary data, broken out by market size and station affiliation. While average news expenses fall as market size falls, that appears to the result of heavy spending by a few stations in large markets; the spending by the median station does not appear to vary significantly by market size for the top 120 markets. It is more likely that in large markets with many stations some stations focus on news programming and others entirely forgo it. In the top 60 markets, more than one-quarter of television stations have no expenses for news programming. Local stations affiliated with ABC, CBS, and NBC are more likely to spend on news programming than stations affiliated with other networks or independent stations. More than half of the CW and MTN affiliates do not have any expenses associated with news programming.

Television stations are taking cost-cutting actions. Even as the quantity of news programming has been increasing, station news staffs, in aggregate, fell by 400 positions in 2009 (primarily in the large markets), after a loss of 1,200 positions in 2008.[78] Almost two-thirds of TV news directors reported budget cuts and staff cuts in 2009.[79] Although hard statistics are not available, the *Pew Study* reported many instances of stations laying off experienced and highly paid staffers, including anchors, on-air reporters, and sportscasters, and of salary reductions, as well as anecdotal evidence of use of part-time journalists and freelance and per-diem staffers to gather and produce the news.[80] It is important to recognize, however, that some of the staffing reductions are the result of technological changes associated with the digitization of television program production that allow stations to use fewer editors and other skilled staff to produce news programming. Most notably, in recent years there has been a trend toward "one-man bands"— reporters equipped with cameras to eliminate the need for separate camera operators.[81]

The *RTDNA/Hofstra Survey* found that, across market size and network affiliation, many television news departments provide content for media platforms other than their own television station, as shown in Table 4. Approximately one-third of all stations produce news programming for another television station in their local market, with this most prevalent among stations in markets 101-150. The survey explicitly asked about mobile devices for the first time in 2010—in past years it was included under "other"—and yet more than 40% of local stations make their programming available on mobile devices. As is the case for broadcast networks, producing content for multiple platforms allows stations to exploit economies of scope by sharing the high fixed costs associated with newsgathering across those outlets.

Table 3. The News Expenses of Local Television Stations, 2009

	Average Station News Expenses	News Expenses as Percentage of Station Expenses (Average)	News Expenses for Station at 25th Percentile	News Expenses for Station at 50th Percentile (Median)
All Stations	$2,662,766	23.9%	$47,420	$1,188,758
Markets 1-10	$9,332,778	24.4%	$0	$2,675,000
Markets 11-20	$4,886,718	24.9%	$0	$1,026,153
Markets 21-30	$4,286,410	25.5%	$0	$4,999,000
Markets 31-40	$3,132,172	25.2%	$0	$2,612,343
Markets 41-50	$2,350,654	23.0%	$0	$1,010,444
Markets 51-60	$1,971,951	22.3%	$0	$1,341,376
Markets 61-70	$1,977,246	22.5%	$246,698	$2,371,189
Markets 71-80	$1,747,958	25.1%	$32,563	$1,414,020
Markets 81-90	$1,640,597	24.4%	$681,065	$1,741,917
Markets 91-100	$1,582,148	22.7%	$812,904	$1,726,716
Markets 101-110	$1,298,857	20.2%	$480,234	$1,542,042
Markets 111-120	$1,342,884	22.1%	$800,347	$1,658,677
Markets 121-130	$1,077,062	21.5%	$338,343	$1,210,838
Markets 131-150	$1,028,904	22.7%	$340,618	$1,085,758
Markets 151-175	$717,734	21.2%	$35,082	$846,120
Markets 176+	$488,507	17.2%	$50,453	$397,721
All ABC, CBS, and NBC affiliates	$3,862,522	28.2%	$1,144,873	$2,011,276
All FOX affiliates	$1,190,646	16.0%	$115,348	$532,370
All CW affiliates	$1,379,084	10.8%	$0	$0
All MNTV affiliates	$534,684	9.0%	$0	$0
All Spanish Language Stations	$771,952	15.5%	$0	$154,182
All Independent Stations	$2,522,099	16.5%	$0	$81,306

Source: National Association of Broadcasters, 2009 Television Financial Report: Station Revenue, Expenses and Profit, various tables.

Table 4. Percentage of Television News Departments Providing Content to Other Media—2010

	Another Local TV Station	TV in Another Market	Cable TV Channel	Local Radio	Website not your own	Mobile Device	Other
All TV Stations	32.7%	13.8%	10.7%	52.0%	13.8%	44.9%	13.8%
Big Four Affiliates	33.5%	12.5%	9.7%	52.8%	12.5%	46.0%	14.8%
Other Commercial Stations	23.5%	23.5%	23.5%	35.3%	29.4%	41.2%	5.9%
Markets 1-25	38.9%	13.9%	16.7%	44.4%	22.2%	55.6%	19.4%
Markets 26-50	22.7%	18.2%	0.0%	63.6%	9.1%	50.0%	9.1%
Markets 51-100	26.2%	13.1%	11.5%	49.2%	14.8%	50.8%	21.3%
Markets 101-150	45.2%	11.9%	4.8%	59.5%	7.1%	31.0%	4.8%
Markets 151+	28.6%	14.3%	17.1%	48.6%	14.3%	37.1%	8.6%

Source: *2010 RTDNA/Hofstra Survey*, Section on How the Business of TV News is Changing, unpaginated.

In addition to producing news programming for others, more than 60% of all stations are involved with some sort of cooperative newsgathering or coverage agreement, as shown in Table 5, which reproduces a table from the *RTDNA/Hofstra Survey*. More than a quarter of those stations that do not currently participate in a cooperative arrangement indicate they are planning or discussing one. The *RTDNA/Hofstra* Survey authors noted, "Interestingly, stations in smaller markets are a little less likely to be involved in cooperative agreements than stations in larger markets." Also, smaller stations are less likely to participate in cooperative agreements than larger ones.

Stations also are increasingly using their newsgathering and news program production capabilities to provide news programming for the non-primary video streams made possible by the digital transition. More than 50% of the news directors responding to the *2010 RTDNA/Hofstra Survey* oversee at least some portion of the programming on a secondary video stream.[82] Programming ranges from all-news formats to weather channels to entertainment programming that includes some local news programming in the schedule. Almost half the news directors that do not currently oversee news programming on their station's non-primary video streams expect to do so sometime in the future.

Station websites are nearly universal; only one responding station in the *RTDNA/Hofstra Survey* did not have one.[83] Only the stations with the smallest news staffs do not include local news on their websites. Nearly all station websites provide news videos, 35.9% provide live newscasts, and 37.5% provide recorded newscasts. On average, station websites have 2.8 full-time and 4.5 part-time staffers. Respondents for 70% of the stations indicated that employees with other responsibilities also help with the website.

Table 5. Percentage of TV Stations with Cooperative Newsgathering or Coverage Agreements with Other Media Outlets (excluding local or nearby TV stations for which station produces news programming)

	Another TV Station	Local Newspaper	Local Radio Station	Other	No
All Stations	23.6%	23.6%	27.7%	4.0%	38.6%
Markets 1-25	41.3%	22.2%	22.2%	9.5%	30.2%
Markets 26-50	22.7%	27.3%	22.7%	2.3%	38.6%
Markets 51-100	18.5%	33.7%	33.7%	6.5%	30.4%
Markets 101-150	13.1%	19.0%	29.8%	0.0%	50.0%
Markets 151+	28.6%	14.3%	25.4%	1.6%	42.9%

Source: *2010 RTDNA/Hofstra Survey,* Section on How the Business of TV News is Changing, unpaginated.

Increasingly, broadcast television stations are pursuing a business model that focuses on local content distributed over the air and on websites to strengthen local brand identity. SNL Kagan has monitored the growing "hyper-localism" strategy of combining the capabilities of stations and their websites to create a strong local brand identity:

> In recent years, local content has been a cornerstone of broadcasters' strategies. Nowadays broadcasters have learned to tailor this local content specifically for the Web to take advantage of local audiences via platforms outside of television. In addition, many broadcasters are engaging in major M&A deals to grow their local content online and increase their market share among competitors.[84]

This strategy may be further strengthened by broadcast-based mobile digital television. An ongoing test by the Open Mobile Video Coalition in

Washington DC has revealed local news to be the most popular type of programming, with spikes in viewing during weather and public safety emergencies.[85]

As a result of the expanded use of station newsgathering capabilities, two-thirds of news directors indicated their station employs a three-screen—on air, online, and mobile—approach to news.[86] More than one-third of all stations are either producing programming for other stations or media outlets or actively involved in cooperative newsgathering or coverage arrangements with other stations or media outlets. These strategies appear to allow many stations to use news programming to compete successfully in the market, even as advertising revenues are challenged.

As the *Pew Study* found: "It comes down to a simple cost-benefit analysis. For most stations, producing more local content is cheaper than paying the fees stations [must pay] for syndication programming."[87]

The various survey results presented in this report suggest that local news programming does not appear to be threatened by market forces currently at play. However, that general conclusion may not reflect the situation facing an individual station when it is making news programming decisions. As a reminder, almost 40% of all broadcast television stations—primarily in large markets—do not broadcast any local news programming. If these stations were to offer original local news programming, that programming would not necessarily increase station profits, particularly for stations with weak brand identify and no existing audience for local news programming.

Local Newsgathering and Programming: Issues for Congress

"Duopolies"

Table 4 and Table 5 show how common it has become, on the local level, for broadcast television stations to cooperate with other media outlets. About one-third of all stations that have their own news departments produce original news programming for other stations in their local market. Also, almost one-quarter of all stations that produce news programming participate in cooperative newsgathering or coverage agreements with other television stations in their local market and a similar number participate in cooperative newsgathering or coverage agreements with newspapers in their local market. As long as each of the stations and newspapers involved retains control of its own programming decisions and operations, such arrangements are not subject to the restrictions in the FCC's local television multiple ownership rule or newspaper-television cross-ownership rule. But the NAB and Newspaper Association of America claim that the sharing activities currently allowed by the rules do not provide sufficient financial relief and seek to reduce or eliminate current restrictions on local television and local television-newspaper mergers.[88]

It is not possible to determine from the data in Table 4 and Table 5 the extent to which the cooperative news activities were fostered by earlier changes in FCC ownership rules—and, if so, whether these rule changes led to more or to less independent news programming. In 1999, the FCC revised its local television ownership rules to allow common ownership of two stations in a market that has at least eight competing television owners, as long as one of the two stations is not among the top four in ratings. (This was intended to prohibit a single entity from operating in a local market two stations that were affiliated with major broadcast networks. Typically, the top-rated stations in a market are affiliated with national news networks. These high ratings often, but not always, hold for local news programming as well as for entertainment programming.) The rule change allows a single entity—including a large station group—to own two stations in a single large market. In the broadcast industry this is often referred to as a "duopoly," although that is not the usual meaning of the word duopoly.[89]

Stations affiliated with major news networks are more likely than other stations to originate local news programming and also are likely to offer more local news programming than other stations (see Table 3). If a duopoly consists of a top-four station and a non-top-four station, the latter is less likely than the former to have been originating its own news programming prior to acquisition. Data are not publicly available to determine whether the rule change allowing duopoly ownership affected the amount of independently produced local news programming in the local market (for example, by one of the stations discontinuing its own news production and letting its partner station produce news programming for it; or by a station producing news programming for its duopoly partner station that had not previously broadcast any news; or by the two stations entering into a cooperative newsgathering or coverage agreement that affects the total amount of news programming aired). It would be informative if the FCC were to collect relevant data and investigate the impact of duopolies on local news programming in the current quadrennial review of its local television multiple ownership rule. This might provide insights on whether the current bright line rule should be retained, eliminated, or modified to allow for the case-by-case determination of whether a proposed combination would be in the public interest.

Local Marketing Agreements (LMAs) or "Virtual Duopolies"

In addition to the shared newsgathering activities, since the 1980s some station owners have entered into various contractual arrangements—local marketing agreements (LMAs), management service arrangements, shared services agreements, and joint operating agreements— whereby one station in a market performs sales, marketing, and operational functions for another station in the market.[90] In 1999 the FCC formally allowed these arrangements, sometimes referred to as "virtual duopolies," as long as the controlling station does not program more than 15% of the programming day of the other station. It is possible for one station to produce or co-produce news programming for its contractual partner without exceeding that 15% ceiling.

In some cases, the two parties to a virtual duopoly arrangement are fully independent. In other cases, it appears that a station group has been actively involved in the creation and financing of a separate entity that has acquired stations in markets in which the station group already owns a station, and then entered into LMAs or similar arrangements with that new entity such that the station group effectively controls more than one station in the market. For example, Cunningham Broadcasting Corporation owns six television stations, all of which are operated by Sinclair Broadcast Group under LMAs. It appears that most Cunningham stock is or has been controlled by trusts in the name of members of the family of the Sinclair founder, who concurrently owned controlling shares of Sinclair.[91] Similarly, Mission Broadcasting, Inc. owns 15 broadcast television stations that all are managed by Nexstar Broadcasting Group through LMAs. Nexstar and Mission have a very close financial relationship. In April, they were co-issuers of $325 million in senior secured second lien notes due in 2017.[92] Sometimes these LMAs exist in markets where a formal duopoly would not be allowed.

Data are not available to determine whether the amount of independently produced local news programming in the local market has changed as a result of these LMA or similar contractual relationships. If the FCC were to collect relevant data and investigate the impact of LMAs on local news programming in the current quadrennial review of its local television multiple ownership rule it might provide insights on whether the rule, which does not address LMAs, should be retained or modified to address LMAs and, if the latter, whether a bright line rule or a case-by-case review would be the best way to serve the public interest.

LMAs and similar arrangements allow broadcasters to exploit potential efficiencies from running multiple stations in a market, which could foster competition, diversity of voices, and localism if the cost savings strengthen the financial viability of marginal stations and foster programming of local interest. But LMAs also have been criticized by multichannel video programming distributors, especially small cable operators, for allegedly enabling broadcasters to raise retransmission consent fees. According to the American Cable Association (ACA),[93] when a single broadcaster can jointly negotiate retransmission consent agreements for two local stations in a market, especially if both stations are affiliated with one of the Big Four broadcast networks that offer "must have" programming, it has great leverage when dealing with a small cable operator. Three small MVPDs that are members of the ACA—Cable America, USA Companies, and Pioneer Telephone Cooperative—filed letters in the FCC's retransmission consent proceeding alleging that when they have negotiated with broadcasters that jointly represented two "Big 4" stations in a market, the average subscriber fee they paid was 161%, 133%, and 30% higher, respectively, than the average fee that they paid to separately controlled Big 4 stations.[94] The cable operators argue that they must pass through these higher payments to their subscribers. Thus, at a time when broadcasters are actively promoting their need for higher retransmission consent payments to be able to support local newsgathering, produce original local news programming, and bid for major sports programming, cable and satellite distributors are pushing back, arguing that these increased retransmission consent fees will result in higher cable and satellite subscriber bills.

LMAs also tend to strengthen the position of the affiliated stations when negotiating advertising rates with local merchants. Rather than competing with one another, the two stations can act as a single entity to maximize their joint profits. Advertisers may benefit by obtaining access to two separate audiences through a single contractual negotiation, and may even be able to obtain volume discounts by advertising on two stations rather than one, but they are likely to be at a disadvantage relative to negotiating with two independent, competing stations. This is especially the case if the two stations are major network affiliates and thus among the small number of media outlets capable of reaching a wide audience. This likely will increase the advertising revenues of the virtual duopoly stations, which would increase the level of resources potentially available to them for local news programming. But this might come at the expense of advertising revenues available to non-duopoly stations in the market. This question might be answered if the FCC conducted a study of the impact of virtual duopolies on retransmission consent fees, on the local advertising market and, indirectly, on the level of local news programming in the current quadrennial review of its local television multiple ownership rule.

Television-Newspaper Combinations

The market pressures on local television stations to develop multiple program platforms to support their newsgathering and news program production facilities and to participate in cooperative newsgathering or coverage agreements with other media outlets in their local markets are also being felt by local newspapers. Some of these newspapers may have an even more urgent need than local television stations to share with or combine with another local newsgathering organization.[95] Under the current rule, a proposed combination between a television station and a major daily newspaper is presumed inconsistent with the public interest if the combination is not in one of the 20 largest markets, or if the television station is among the four highest-rated stations in the market, or if after the transaction there were no longer at least eight independently owned and operating major media voices in the market. However, that presumption is reversed— it is presumptively in the public interest—if the proposed merger includes a newspaper or television station that has failed or is failing. In addition, one of the factors considered to confirm or rebut the positive or negative public interest presumption is the financial condition of the newspaper and television station and, if either is in financial distress, the proposed owner's commitment to invest significantly in newsroom operations.

In general, newspapers in smaller markets are in less distress than major daily newspapers in large markets. CRS analyst Suzanne Kirchhoff found: "Smaller papers are in a better financial position than large dailies for several reasons. Smaller papers are less dependent on classified ads, operate in less complex markets, and tend to be closer to their readers and advertisers than large dailies."[96]

This would suggest that the current FCC cross-ownership rule—which presumes a proposed newspaper-television merger in the largest 20 markets to be consistent with the public interest, but presumes such a merger in a smaller market to be inconsistent with the public interest— reflects current market conditions. At the same time, the current rule takes into account financial distress when addressing proposed mergers in smaller markets, though the burden remains on the entities proposing the merger to make a public interest case.

The ability to claim financial distress raises an interesting policy issue. In the past decade, a number of broadcast station groups and newspaper groups expanded through acquisition, using highly leveraged financing that left them with sizable debt loads.[97] When television and television advertising revenues fell during the economic downturn, some of these companies did not generate enough revenues to cover their interest expenses in addition to their operating expenses; the companies had positive cash flows, but pre-tax losses. In some cases, they were in breach of financing agreements or covenants with their lenders or at least risked being in breach. What if such a newspaper or television station group chose to use this financial distress as the basis for justifying an otherwise prohibited merger? If the station or newspaper enjoyed a positive cash flow, but experienced distress in significant part as a result of choosing a risky method of financing expansion, could that distress provide the basis for rebutting a negative public interest presumption against allowing that station or newspaper to merge with another local station? Or would the proper test be whether the station or newspaper would be in stress even if relieved of its heavy debt load, which could happen if there were a restructuring with the lenders assuming equity positions?

Market Forces Affecting Broadcast Carriage of Major Sports Events

Unique Supply-Side and Demand-Side Characteristics

Unique supply-side and demand-side characteristics of major professional, intercollegiate, and Olympic sports programming create market forces that are very different than those for other types of programming, especially news programming. On the supply side, major sports programming presents unique, time-sensitive events for which there is no close substitute programming. A competing programmer cannot create copycat programming, the way a popular scripted or unscripted show could spur imitators that potentially could create an even more highly demanded product.[98] Moreover, with major sports programming there is a well-defined entity—a professional sports league, a major intercollegiate conference, or an International Olympic Committee—that controls access to the events that provide the basis for the programming and that typically retains control of the programming itself. This is in sharp contrast with news programming, which primarily covers events to which competing programmers have roughly the same access. As a result of their control over access to the sports events, these sports entities are able to negotiate with programmers and distributors from a position of strength.[99]

On the demand side, there are several distinctive characteristics of the audience for major sports programming. First, a minority of U.S. households have such a high intensity of demand for the programming that they consider it "must have" programming. They are likely to decide whether to rely on free over-the-air broadcast service or to subscribe to a pay multichannel video service—and which such service to subscribe to—based upon which option gives them the most favorable access to the desired sports programming.[100] The behavior of this small minority of total television households affects the sports programming decisions of pay multichannel video distributors. At the same time, there is a much larger total audience for major sports programming.[101] Individual households in that mass audience are less likely to base their decisions to subscribe or not subscribe to a pay service on the availability of that programming and are unlikely to be willing to pay specifically for that programming. Both the high intensity audience and the mass audience include men who watch very limited amounts of other types of programming and therefore represent a unique target for advertisers. The entity that controls the sports programming will take into account these demand characteristics—mass appeal, high-intensity appeal, and unique appeal—when constructing a profit-maximizing distribution strategy.

Sports Programming is of High Value to Video Distributors Even if Not Profitable

Program networks and multichannel video providers have long utilized major sports programming to promote their own brand identities and major sports entities have been able to take advantage of that to generate competitive bidding for the rights to their sports programming.

- In 1987, ESPN obtained rights to offer ESPN Sunday Night Football, on the condition that it simulcast the games on local broadcast stations in the participating markets. Although ESPN had carried National Basketball Association games in 1982-1984, its acquisition of rights to National Football League games was viewed as an important step in attaining credibility as a major distributor of sports programming.
- In 1993, when the Fox Broadcasting Company's network had neither the full schedule of programs nor the geographic reach of the three established broadcast networks (ABC, CBS, and NBC), it successfully bid $1.58 billion for four years of programming rights to the National Football Conference of the NFL. At the time, the Fox network had neither a sports division nor a news division, but by winning the rights

and then attracting key on-air talent away from CBS (which lost the rights), it created credibility that allowed it both to purchase additional owned and operated stations for its network and to attract additional affiliates.

- More recently, DirecTV has obtained the exclusive rights to the NFL Sunday Ticket package of "out of market" games[102] and other sports programming as part of a strategy to brand itself as the premier provider of sports programming.

As the proliferation of cable networks led to audience fragmentation, major sports programming that attracted a large audience offered new value to the broadcast or cable network that won the programming rights—a means to market the other programs on the network to the large sports viewing audience. It became common for the game announcers to market other network programs by identifying the stars of those programs that were in attendance at the sports event being aired. When the network with the rights to the sports programming is owned by a major programmer with multiple broadcast and/or cable networks, the major sports programming also can be utilized to cross-market the programming aired on those other networks. As a result, the bids for sports programming rights increased even though the sports programming on its own did not generate enough revenues to directly cover costs.[103]

One reason why bids increased is that, with the successful entry of cable networks, there were more entities able to bid. By 1998, the most successful cable networks, such as TBS, TNT, and ESPN, each had achieved penetration into more than 75 million cable and satellite households; by 2007, they were received by more than 95 million households.[104] Although the cable networks generated significantly lower advertising revenues than the broadcast networks, they generated very substantial per subscriber fees. In 2003, at a time when broadcast stations were not receiving cash payments from cable operators for retransmission consent rights, ESPN commanded monthly fees of $2.17 per subscriber, TNT $0.77, and TBS $0.30.[105] SNL Kagan estimates that in 2010 ESPN commanded $4.41; TNT commanded $0.96; and TBS $0.48. These popular cable networks all were carried on the cable and satellite tiers reaching the largest number of subscribers.

In 2008, subscriber license fees represented 54.0% of revenues for advertiser-supported cable networks; advertising revenues only represented 42.2% of revenues.[106] In 2009, advertising revenues fell while subscriber fees grew, so today advertising represents an even smaller share of cable network revenues. In contrast, advertising continues to represent the preponderance of broadcast network revenues, in the vicinity of 90%. It is only recently that broadcasters began to collect retransmission consent compensation from multichannel video providers in the form of per subscriber fees. Given that advertising revenues are highly sensitive to the business cycle, and often fall during recessions, broadcast networks rely on a less stable source of revenues than cable networks.

To compensate for this, broadcast networks increasingly are demanding that their local station network affiliates help fund the acquisition of expensive sports (and other) programming, either through direct affiliate payments to the network or by requiring the affiliates to share a portion of their retransmission consent revenues with the network.[107]

The bidding for the programming rights to major sports events occurs years before the events, at which time it is not possible to accurately predict economic conditions in the months leading up to the event, when the bulk of the advertising will be sold. As they bid for rights against cable networks, broadcast networks invariably have less certain projections of the revenues they will have available to support the bid. Thus, in February 2002 News Corp., parent of Fox Entertainment Group, cited the downturn in advertising as the reason why it took a $909 million one-time operating charge for broadcast sports contracts with the National Football League, Major League Baseball, and Nascar auto racing.[108] In 2003, NBC bid $2.2 billion for the rights to the 2010 and 2012 Olympic Games (including $820 million for the 2010 Winter Olympics), a 33% increase over the rights fees for the 2006 and 2008 Olympic Games,[109] but had to attempt to sell advertising during 2009 and early 2010, when the advertising market was at its nadir. NBC has indicated that it lost approximately $200 million on the 2010 Olympics.[110] In comparison, for the 2006 Olympics, when General Motors was the primary advertiser, NBC was able to generate advertising revenues that exceeded rights payments by about $200 million.[111] Many observers believe the 2010 experience will constrain future bids for the programming rights to the Olympics.

The Distribution of Major Sports Programming over Multiple Platforms

Despite the negative financial bottom-line of the 2010 Olympic programming, the strategy employed by NBC is likely to be the model for the future. NBC fully exploited its wide range of distribution outlets—the USA,

MSNBC, CNBC, and Universal HD cable networks, as well as NBCOlympics.com—to provide far more complete coverage of the Olympics than ever before. The NBC broadcast network and its cable networks offered more than 835 hours of sports programming—more than the total number of hours that elapsed between the opening and closing of the Olympics. Multiple events occurred simultaneously and thus could not be covered by a single network, but the combination of broadcast and cable networks was able to provide extensive coverage, including significant amounts of live coverage. The NBC broadcast network did not want to pre-empt so much of its non-Olympics programming that it undermined its other programming, but the 835 hours of Olympic programming could be borne by a combination of networks with less impact on their program schedules. There also was coordinated coverage on a website, NBCOlympics.com. The high fixed costs associated with the rights fees and production facilities were spread across the many NBC-Universal affiliated distribution outlets, and the total quantity of Olympics programming far exceeded that for any previous Olympic Games.[112]

This strategy of using several outlets to cover a complex sports competition also was used in the recent successful bid by CBS and the Turner Cable Networks for the rights to future National Collegiate Athletic Association basketball tournaments (March Madness). Even as the tournament expands to include more teams, and thus encompass more games, the wider reliance on cable networks to supplement the broadcast network will allow all games to be televised. The fixed costs of production will be shared across more hours of programming.

Similarly, the strategy of coordinating television and Internet is being adapted to other sports. A recent GMR Marketing survey of consumers in markets where Comcast offers cable service found that while television remains the most important medium for National Football League fans, "the number of multimedia multi-taskers is growing at a swift pace due to the enormity of fantasy sports,"[113] with 76% of the respondents who identified themselves as NFL fans using the Internet to follow sports at the same time as they are watching sports on television.

The Rise of Cable Networks Owned by Major Sports Entities

There is a major new development in the sports programming market. A number of major sports entities with strong brand identities—leagues such as the National Football League and the Big Ten Intercollegiate Conference, and teams such as the New York Yankees and Baltimore Orioles—have created their own national or regional cable networks and distributed some or all of their sports events exclusively over those networks. This business strategy allows leagues or teams to leverage their control of a limited amount of seasonal "must have" programming, which represents only a relatively small number of hours in a network's annual programming schedule, into a package that can command payment year round. In effect, it ties access to the limited amount of must-have programming to the purchase of lots of less-demanded programming, such as "classic" game re-runs, interviews, pre- and post-game shows, pre-season shows, player drafts, and less popular sports events.

In general, these successfully branded sports entities have insisted that cable and satellite operators place their sports networks on the tier with the largest number of subscribers, in order to generate monthly per subscriber fees from the maximum number of subscribers. This also allows the sports entity to receive payment for the entire year. If the must-have programming were placed on a sports tier, fans could sign up just for the season and then drop the tier. Although cable companies initially resisted the requirement that the sports networks be placed on the tier with the largest number of subscribers—preferring that the sports networks be put in a special sports tier—the sports entities have succeeded in their negotiations by using the threat that those subscribers with a high-intensity of demand will desert the cable or satellite operator who does not carry the sports channel on a basic tier for an operator who does carry it on a basic tier. The sports entities have been able to exploit the fact that their events represent must have programming to a sufficiently large minority of television households to require all households to pay for the programming through the entire year. This is likely to reinforce the long-standing pattern of sports networks accounting for a much larger percentage of cable and satellite operator programming costs than their percentage of cable and satellite subscriber viewing.[114]

The unique supply and demand characteristics of major sports programming appear to have two significant implications for future broadcast carriage of such programming. On one hand, the sports entities control access to the sports events and therefore have the ability to command for themselves most of the economic rents generated by the intensely valued sports programming.[115] Broadcasters may have relatively little control over the largest programming cost—the license fee for the right to broadcast the sports events. On the other hand, broadcast networks are adapting to this by coordinating with other program distributors to compete more effectively when bidding for programming rights and to generate more revenues when they win the rights. At a minimum, broadcast networks are seeking a share of the fees their affiliates are able to command from cable and satellite operators for the retransmission of intensely demanded network-produced major sports programming. In addition, broadcast networks are collaborating with affiliated or independent cable and Internet outlets to develop multi-platform distribution strategies that generate greater revenues from the programming rights. These strategies cede a greater portion of sports programming to cable networks than in the past, but allow broadcasters to retain a share of that programming.

Sports Programming: Issues for Congress

Given that the entities that control major sports events have much greater control over the production and distribution of major sports programming than do broadcasters, existing regulatory rules that are directed at broadcasters—whether retained or modified—are unlikely to have much impact on broadcast provision of major sports programming. If Congress is concerned about how the public can access major sports programming—for example, if it is concerned that such programming be available over the air without subscription or that cable subscribers not be required to pay for such programming even if they do not view it—then it will need to examine sports entities as well as the broadcast and cable networks. Issues might include the program blackout policies of the sports entities, the antitrust laws as they apply to sports entities, and the FCC's program carriage rules applicable to impasses in negotiations between sports entities that own cable networks and multichannel video programming distributors.

- **Blackouts.** Some National Football League teams are not selling out tickets to their home games. They often would prefer to televise those games to hometown fans in order to foster broad-based interest in and support for the team, but league rules impose a blackout on games when tickets are not sold out 72 hours in advance of the game. The NFL has made the determination that it is beneficial to the league to blackout those games, even if the home team and its fans are harmed. NFL officials are concerned that if a stadium is only half full, it also harms the quality of the televised program because viewers prefer to see the frenzy of a full stadium of fans cheering on their team.[116] In many cases, the team's stadium was directly or indirectly funded by a governmental entity, and taxpayers may feel deceived if they have borne some of the costs of the stadium but are denied television access to the games. If blackouts were prohibited and as a result stadium attendance fell because some fans chose to watch the games on television rather than at the stadium, team revenues and local tax revenues from ticket sales and food and parking concessions could fall. In the extreme case, a team's ability to make lease payments to the stadium owner could be impaired. In most situations, however, the fall in stadium attendance would be too small for the impact on team revenues and tax revenues to be substantial. Is market intervention justified if a league's joint profit maximizing behavior is harmful to a team and that team's community and, if so, can an intervention be devised that is not itself harmful?

- **Bundling.** Consider a major sports entity with programming that is so intensely valued by a sufficiently large minority of households (who would threaten to abandon a multichannel video provider that does not carry that programming) that the sports entity can insist that all multichannel video providers carry that programming on their tier with the largest subscriber base, even if the multichannel video provider would prefer to carry it on a special sports tier. This may maximize revenues for the sports entity but at the expense of a majority of multichannel video subscribers who do not watch that programming but must bear some of its cost. Is market intervention justified if it can be shown that there is significant loss in consumer welfare as a result and, if so, can an intervention be devised that is not itself harmful?[117]
- **"Free" broadcasts.** With more than 85% of all U.S. households subscribing to a multichannel video provider, and more than 90% of those households purchasing more than just a basic package of programming, is there any public policy basis for market intervention to keep sports programming on "free" broadcast television? If so, can intervention be devised that is not itself harmful?

End Notes

[1] In the Matter of 2010 Quadrennial Regulatory Review—Review of the Commission's Broadcast Ownership Rules and Other Rules Adopted Pursuant to Section 202 of the Telecommunications Act of 1996, MB Docket No. 09-182, Notice of Inquiry, adopted and released May 25, 2010.

[2] For a detailed discussion of the FCC's broadcast media ownership rules, see CRS Report RL34416, The FCC's Broadcast Media Ownership Rules, by Charles B. Goldfarb.

[3] In the Matter of 2010 Quadrennial Regulatory Review—Review of the Commission's Broadcast Ownership Rules and Other Rules Adopted Pursuant to Section 202 of the Telecommunications Act of 1996, MB Docket No. 09-182, Comments of the National Association of Broadcasters, July 12, 2010, at pp. i-iv.

[4] In the Matter of 2010 Quadrennial Regulatory Review—Review of the Commission's Broadcast Ownership Rules and Other Rules Adopted Pursuant to Section 202 of the Telecommunications Act of 1996, MB Docket No. 09-182, Comments of the National Association of Black Owned Broadcasters, July 12, 2010, at p. iii.

[5] In statutes and FCC rules, these firms are formally referred to as "multichannel video programming distributors" or "MVPDs." For ease of presentation, in this report they will be referred to as "multichannel video distributors."

[6] 47 C.F.R. §§ 76.64-65, which implement statutory provisions that were added to the Communications Act by the Cable Television Consumer Protection and Competition Act of 1992, P.L. 102-385, 106 Stat. 1460 (1992). For a brief discussion of the retransmission consent rules, see the section entitled "Broadcast Networks and Their Affiliated Local Broadcast Stations" in CRS Report R41063, The Proposed Comcast-NBC Universal Combination: How It Might Affect the Video Market, by Charles B. Goldfarb.

[7] Public Notice, Media Bureau Seeks Comment on a Petition for Rulemaking to Amend the Commission's Rules Governing Retransmission Consent, DA 10-474 , Media Bureau, Federal Communications Commission, released March 19, 2010.

[8] In the Matter of Petition for Rulemaking to Amend the Commission's Rules Governing Retransmission Consent, Opposition of the Broadcaster Associations, May 18, 2010.

[9] Matthieu Coppet, et. al., "Broadcast Rally: When Will the Music Stop," UBS Investment Research, UBS Global I/0: Global Media, March 15, 2010.

[10] See, for example, Pew Project for Excellence in Journalism, State of the News Media 2010 (Pew Study), Section on Local Television Economics, unpaginated, available at http://www.stateofthemedia.org/2010/, viewed on August 23, 2010, which states that despite the fall in advertising revenues, in 2010 they still will represent 91% of local broadcast station revenues.

[11] See SNL Kagan, Economics of Basic Cable Networks, 2009 Edition, at p. 2.

[12] These include Magna Global, BIA/Kelsey, Veronis Suhler Stevenson, and SNL Kagan, as well as analysts at brokerage and other financial firms, such as UBS.

[13] Magna Global can be contacted at http://www.magnagloval.com/?action=register.

[14] It is likely, however, that in at least some cases the broadcast network will be part of the same parent company as the cable network.

[15] Interestingly, none of the forecasters projected the high level of revenue growth that a number of broadcast station owners have been reporting for the second quarter of 2010. Although most forecasters projected rates of growth (after the 2008-2009 decline) in the vicinity of 5%, most broadcast television station groups have reported double digit revenue growth in the second quarter of 2010, with some reporting growth approaching 40%. For example, Fisher Television reported a 37% increase in television revenues over the same quarter last year (see Michael Malone, "Fisher Television Revenue Up 37%," Broadcasting & Cable, August 9, 2010) and E.W. Scripps reported a 26% increase in advertising revenues at its stations in the second quarter (see David Goetzl, "E.W. Scripps: Broadcast Enjoys Ad Hike, Newspapers Dip," Media Post News, August 9, 2010). This discrepancy between the forecasts of revenue growth and the actual second quarter 2010 results can be at least

partially explained: forecasters do not attempt to estimate quarterly rates of growth, but rather try to project an average rate of growth over a longer period of time. To do this, they must factor in, for example, the risk of a double-dip recession beginning later in the year that would not affect second quarter 2010 revenue levels, but could affect later revenue levels.

[16] SNL Kagan is a research firm specializing in media and communications whose data are widely used in the industry. The SNL Kagan estimates for 2009, 2010, and 2011 were reported in Michael Malone, "Local Broadcasters Bullish at SNL Conference," Broadcasting & Cable, June 26, 2010. The SNL Kagan projection for 2016 was reported in Steven C. Salop, Tasneem Chipty, Martin DeStefano, Serge X. Moresi, and John R. Woodbury, "Economic Analysis of Broadcasters' Brinkmanship and Bargaining Advantage in Retransmission Consent Negotiations," a study prepared at the request of Time Warner Cable, June 3, 2010, at p. 18, Figure 4. The study is available at http://www.american televisionalliance.org under "Broadcaster Brinksmanship."

[17] See, for example, Glen Dickson, "Mobile DTV Joint Venture Outlines Leadership: Fox, NBC execs to handle product development, distribution for Mobile Content Venture," Broadcasting & Cable, June 8, 2010.

[18] See "Broadcast Note," Communications Daily, September 9, 2010.

[19] See, for example, William Kidd, "Bandwidth Caps Are a Barrier to Emerging Internet TV Competition," iSuppli Applied Market Intelligence, July 20, 2010, available at [http://www.isuppli.com/Home-and-Consumer-Electronics/ News/Pages/Bandwidth-Caps-Area-Barrier-to-Emerging-Internet-TV-Competition.aspx.], viewed on August 11, 2010.

[20] See, for example, Mary Collins, "Turning Subchannels Into Revenue," TVNewsCheck, July 30, 2010, available at http://www.tvnewscheck.com/article/2010/07/30/44016/turning-subchannels-into-revenue, viewed on September 14, 2010, citing a study performed by International Media Advisors and Bortz Media & Sports Group.

[21] See, for example, Matthieu Coppet, et al, "3D TV is here to stay," UBS Investment Research, UBS Global I/O: 3D TV and Media, January 15, 2010, which suggests that consumers are most likely to obtain 3D TV service from the multichannel video distributors.

[22] According to a March 2010 Pew Internet & American Life Project report, entitled "Understanding the Participatory News Consumer," at p. 5, available at http://www.pewinternet.org/Reports/2010/Online-News.aspx?5=1#, "78% of Americans say they get news [on a typical day] from a local TV station; 73% say they get news from a national network such as CBS or cable TV station such as CNN or Fox News." Although most Americans now get their news information from multiple sources, no other source is as widely used as television.

[23] See, for example, Pew Study, Section on Network TV Audience, at p. 4, which explains that the lowest-rated broadcast network newscast (CBS Evening News with Katie Couric) had an average nightly audience of 5.9 million viewers for the year, while the most popular prime time cable news program (Fox News Channel's O'Reilly Factor) had an audience of 3.3 million.

[24] Pew Study, Section on Network TV Audience, at pp. 1-5.

[25] Pew Study, Section on Network TV Audience, at p. 5.

[26] These actions are discussed below in the section entitled "Market Forces Affecting Broadcast News Networks."

[27] Pew Study, Section on Local TV Audience, at pp. 1-5.

[28] See footnote 22 above.

[29] See Bob Pepper, 2010 RTDNA/Hofstra Staffing & Profitability Survey (RTDNA/Hofstra Survey), Section on TV Staffing and News 2010, unpaginated, available at [http://rtdna.org/pages/media_items/2010-rtdnahofstra-staffingprofitability-survey—full-data1944.php], viewed on July 22, 2010. According to the survey, the average amount of original local news programming broadcast by those stations that produce their own programming increased to 5.0 hours per weekday in 2009 from 4.7 hours in 2008.

[30] These actions are discussed below in the section entitled "Market Forces Affecting Local Broadcast Station News."

[31] See RTDNA/Hofstra Survey, Section on How the Business of TV News is Changing, unpaginated.

[32] See, for example, "Mass Media Notes," Communications Daily, April 22, 2010.

[33] Mark J. Prak, David Kushner, and Eric M. David, "The Economics Realities of Local Television News—2010: A Report for the National Association of Broadcasters" (NAB Survey), In the Matter of 2010 Quadrennial Regulatory Review—Review of the Commission's Broadcast Ownership Rules and Other Rules Adopted Pursuant to Section 202 of the Telecommunications Act of 1996, Comments of the National Association of Broadcasters, Attachment B, July 12, 2010, at p. 30.

[34] In reviewing the financial data in this report specific to newsgathering and news (and sports) program production, please note that neither broadcast networks nor local broadcast stations are required to report revenues, costs, and profits at that level of disaggregation. The data presented therefore either are indirect estimates prepared by researchers or the aggregated responses of individual news directors to survey questions, where the respondents may not all be basing their responses on the same set of definitions and assumptions. As a result, it generally is more useful to rely on these data to identify trends over time than to provide point estimates of revenues, costs, or profits.

[35] See Pew Study, Section on Network Television, and Marc Gunther, "The Transformation of Network News: How Profitability Has Moved Networks Out of Hard News," Nieman Reports Special Issue 1999 (Nieman Report), available at http://www.nieman.harvard.edu/reportsitem.aspx?id=102153, viewed on August 16, 2010.

[36] The Fox Broadcasting Company provides its affiliates the one-hour weekly public affairs program Fox News Sunday. It also provides coverage of political events such as the State of the Union address and national election coverage and breaking news, with programming produced by, but separate from the programming provided on, the cable Fox News Channel. Also, the Fox News Edge service provides national and international news reports for local Fox affiliates to use in their own newscasts.

[37] See footnote 24 above.

[38] Nieman Report, at p. 1. The following discussion of network news profitability presents estimates made by industry analysts who had to make many explicit or implicit assumptions about how much of the parent company's overhead and other costs should be assigned to the broadcast news network. It is not possible to verify the findings of these analysts.

[39] According to the 1999 Nieman Report, NBC News earned more than $200 million in 1998 and its profits had grown steadily during the decade. The Pew Study estimates 2008 NBC News pre-tax profits of $400 million, although about two-thirds of its revenues came from the cable networks MSNBC and CNBC.

[40] The 1999 Nieman Report estimated that ABC News earned $55 million in 1998, down from a peak of $110 million two years earlier. It also estimated that CBS News earned only $15 million in 1998, but that those earnings were growing. The 2010 Pew Study estimated that the ABC news division may have generated a small operating profit in 2009 on total

revenues of approximately $600 million. It also estimated that CBS News lost money in 2009 on revenues of about $400 million.

[41] Pew Study, Section on Network TV Audience, at p. 2, citing Nielsen Media Research data.

[42] See footnote 26 above.

[43] Pew Study, Section on Network TV Audience, at pp. 6-8, citing Nielsen Media Research data.

[44] The following discussion of the 1990s is based on the presentation in the Nieman Report.

[45] The following discussion of 2010 is based on the presentation in the Pew Study.

[46] Nieman Report, Section on Economics of Network News, unpaginated.

[47] Pew Study, Summary Essay in Section on Network TV, at pp. 1-2.

[48] Such speculation is not new. According to the Nieman Report, in 1998 ABC, NBC, and CBS each had discussions with CNN about sharing staff and bureaus outside the United States.

[49] The FCC's Dual Network Ownership rules prohibits a merger among the "top four" broadcast networks—ABC, CBS, FOX, and NBC—but does not relate to cable networks. The FCC's National Television Ownership Rule allows a broadcast network to own and operate local broadcast stations that reach, in total, up to 39% of U.S. television households, but does not relate to the reach of co-owned cable networks.

[50] See, for example, CRS Report R40700, The U.S. Newspaper Industry in Transition, by Suzanne M. Kirchhoff.

[51] The broadcast-newspaper cross-ownership rule had been stayed by the Third Circuit Court of Appeals while it is being appealed, but the court has vacated its stay. See John Eggerton, "Third Circuit Lifts Stay on Media Ownership Rules," Broadcasting & Cable, March 23, 2010.

[52] Pew Study, Section on Cable TV Audience, unpaginated, based on Nielsen Media Research data.

[53] Ibid. Interestingly, although Fox prime-time programs have the highest ratings among the three networks, CNN continues to have the highest cumulative audience.

[54] Pew Study, Section on Cable TV Economics, unpaginated.

[55] Ibid.

[56] Pew Study, Section on Newspaper Audience, unpaginated, citing Editor and Publisher Yearbook data.

[57] Ibid, citing Audit Bureau of Circulations and Newspaper Association of America data.

[58] Pew Study, Section on Newspaper Economics, unpaginated, citing data from the Newspaper Association of America and newspaper industry analyst Rick Edmonds.

[59] Ibid.

[60] This may be unique programming produced exclusively for the station, the rebroadcast of programming aired by the producing station, or a combination of these.

[61] These aggregate figures were constructed by Bob Pepper, the Lawrence Stessin Distinguished Professor of Journalism and chairman of the Department of Journalism, Media Studies, and Public Relations at Hofstra University. He is the primary researcher and author of the annual RTDNA/Hofstra Staffing and Profitability Survey of broadcast station news directors cited earlier in this report that is widely recognized within the industry. Professor Pepper found that currently 737 non-PBS stations produce local programming, 214 run programming of local interest produced by other stations, and 609 do not air local news programming. He also found that 19 Public Broadcasting System (PBS) stations originate local news programming and 181 do not broadcast local news programming.

[62] 2010 RTDNA/Hofstra Survey, Section on TV Staffing and News 2010, unpaginated. The amount of original local news programming is much lower on weekends, averaging 1.7 hours on Saturdays and 1.6 hours on Sundays.

[63] RTDNA/Hofstra Survey, Section on TV Staffing and News, unpaginated.

[64] Pew Study, Section on Local TV Audience, at pp. 1-5.

[65] National Association of Broadcasters, annual Television Financial Report: Station Revenue, Expenses and Profit, from 1980s through 2010.

[66] Unfortunately, the NAB's annual Television Financial Reports in the 1990s do not include average and percentile data for "all stations in all markets," but they do include tables for many different subsets of television stations and a review of them indicates that, for stations as a whole, the dollar level of profits fell substantially in 1999 and has not recovered to the 1998 level. Through 1997, only a relatively small share of stations had before-tax losses; since 1998, at least one-fourth of all stations have had pre-tax losses.

[67] NAB Survey, Attachment C, at p. 2. The NAB table provides pre-tax station profits for the average station, the station in the 25th percentile for profits, the 50th percentile (the median station), and the 75th percentile for the years 1998-2008, based on data compiled by the NAB is an annual survey of more than 700 stations. Unfortunately, those data are not included in the NAB's annual Television Financial Reports prior to 2002 and therefore it is not possible to reproduce the data the NAB provides for the period 1998-2001 and it is not possible to use the annual Television Financial Reports to extend the overall profits data back earlier than 1998. But a review of the data in the earlier Financial Reports strongly indicates that stations in the mid-1990s earned higher absolute levels of profits than they have subsequently.

[68] Table 1 of the NAB's 2009 Television Financial Report presents "national average" data for all stations, for the overall average, the 25th percentile, 50th percentile (median), and 75th percentile. For both the 25th and 50th percentile, interest expenses are zero. For the 75th percentile the interest expenses are $431,571. Average interest expenses are $772,920. This suggests that some stations have very high debt loads. (There are several reasons why a company might choose a highly leveraged financial structure. The tax code creates an incentive since interest payments are tax deductible for the company but dividend payments (to equity holders) are not. Also, there is no dilution of the equity base if a company chooses to fund expansion (whether internal or by acquisition) by using debt financing. If an owner chooses a highly leveraged financing strategy, it trades off higher equity capital requirements for higher interest expenses.)

[69] The 2009 Television Financial Report presents 2008 operating cash flow data for 85 different categories of television stations (by network affiliation, by size of market, by station revenues, and various combinations of these). Of these 85 categories, the cash flow is negative for the station in the 25th quartile for 10 categories of stations: all affiliate and independent stations in markets 121-130; ABC, CBS, and NBC affiliates with net revenues between $5 million and $6 million; ABC, CBS, and NBC affiliates with net revenues between $4 million and $5 million; ABC, CBS, and NBC affiliates with net revenues between $2 million and $3 million; ABC, CBS, and NBC affiliates with revenues under $2 million; CW affiliates in markets 76+; Spanish language stations in markets 1-25; all independent stations; independent stations in markets 1-25; and independent stations in market 26+. For one category—CW affiliates in markets 76+—at least half the stations had a negative cash flow. Unfortunately, the Television Financial Report does not indicate the number of stations in each of its categories; it is possible that some of these tables are reporting the financial data of only a very few stations.

[70] The format in which the data are presented in the NAB's annual Television Financial Reports does not allow for an analysis of the impact of debt load and interest expenses because it is presented in percentile format. The station in the 25th percentile for pre-tax profits is

unlikely to be the same station that is in the 25th percentile for cash flow, or the station that is in the 25th percentile for interest expenses. As a result, there is no way to use the data as presented by the NAB to determine the impact of debt load on profits.

[71] Cash payments for retransmission consent is a relatively new phenomenon. Major networks provide more "must-have" programming of the sort that can command high cash payments than do non-major networks or independent stations.

[72] This is the price that a religious broadcaster or other non-profit broadcaster would seek to pay to acquire a license from an existing licensee. Stations with positive cash flows would of course command a higher price.

[73] SNL Kagan, Broadcast Investor: Deals & Finance, Number 538, June 29, 2010, at p. 4.

[74] See, for example, Steve Lawson, "How Much is a TV Station worth today?" on the blog Explaining Social Media and Social Networking, February 24, 2010, available at [www.friendlyvoice.com/blog/2010/02/24/how-much-is-a-tv- station-worth-today], viewed on October 12, 2010.

[75] RTDNA/Hofstra Survey, Section on TV News Staffing and Profitability, at p. 4. 71.4% of survey respondents said they did not know how much revenue (or what percentage of station revenue) came from news programming.

[76] NAB Survey, at p. 10.

[77] RTDNA/Hofstra Survey, Section on TV News Staffing and Profitability, at p. 3.

[78] RTDNA/Hofstra Survey, Section on TV News Staffing and Profitability, at p. 1.

[79] RTDNA/Hofstra Survey, Section on TV News Staffing and Profitability, at p. 1.

[80] Pew Study, Section on Local TV News Investment, at pp. 2-3.

[81] RTDNA/Hofstra Survey, Section on One-Man-Bands 2010 Update.

[82] RTDNA/Hofstra Survey, Section on How the Business of TV News is Changing, unpaginated.

[83] This discussion of station websites is based on the presentation in RTDNA/Hofstra Survey, section on "TV and Radio on the Web," unpaginated.

[84] E-mail message from SNL Kagan to clients, August 19, 2010, marketing its scheduled video presentation entitled "Watch: SNL Kagan analyzes the broadcast strategy of tailoring hyper-local content."

[85] "Broadcast," Communications Daily, September 15, 2010.

[86] RTDNA/Hofstra Survey, Section on How the Business of TV News is Changing, unpaginated.

[87] Pew Study, Section on Local TV News Investment, unpaginated.

[88] In the Matter of 2010 Quadrennial Regulatory Review—Review of the Commission's Broadcast Ownership Rules and Other Rules Adopted Pursuant to Section 202 of the Telecommunications Act of 1996, MB Docket No. 09-182, Comments of the National Association of Broadcasters, July 12, 2010, at pp. i-iv, and In the Matter of 2010 Quadrennial Regulatory Review—Review of the Commission's Broadcast Ownership Rules and Other Rules Adopted Pursuant to Section 202 of the Telecommunications Act of 1996, MB Docket No. 09-182, Comments of the Newspaper Association of America, July 12, 2010, at p. 1.

[89] To economists, and in common usage, a "duopoly" is a market situation in which there are only two producers, so each firm consciously takes into account the decisions of the other when making its own business decisions.

[90] This discussion is based on Kim McAvoy, "Virtual Duopolies Coming Under Fire," TVNewsCheck Focus on Washington, June 9, 2010, available at [http://www.tvnews check.com/articles/2010/06/09/ daily2/], viewed on August 31, 2010.

[91] See, for example, Leon Lazaroff, "Media Firm Accused of Dodging FCC Rules, Knight-Ridder, October 17, 2004.

[92] United States Securities and Exchange Commission, Mission Broadcasting Inc. Form 8-K, April 12, 2010.

[93] In the Matter of 2010 Quadrennial Regulatory Review—Review of the Commission's Broadcast Ownership Rules and Other Rules Adopted Pursuant to Section 202 of the Telecommunications Act of 1996, MB Docket No. 09-182, Comments of the American Cable Association, July 12, 2010, at pp. 11-17.

[94] In the Matter of Petition for Rulemaking to Amend the Commission's Rules Governing Retransmission Consent, MB Docket No. 10-71, letter dated May 28, 2010, from Christopher A. Dyrek, executive vice president, Cable America, to Marlene H. Dortsch, Secretary, Federal Communications Commission.; letter dated May 28, 2010, from Christian M. Hilliard, president, USA Companies, to Marlene H. Dortsch; letter dated June 4, 2010, from Scott Ulsaker, manager Pioneer Long Distance, Pioneer Telephone Cooperative, to Marlene H. Dortsch.

[95] See CRS Report R40700, The U.S. Newspaper Industry in Transition, by Suzanne M. Kirchhoff. See also In the Matter of 2010 Quadrennial Regulatory Review—Review of the Commission's Broadcast Ownership Rules and Other Rules Adopted Pursuant to Section 202 of the Telecommunications Act of 1996, MB Docket No. 09-182, Comments of the Newspaper Association of America, July 12, 2010, at p. 3.

[96] CRS Report R40700, The U.S. Newspaper Industry in Transition, by Suzanne M. Kirchhoff.

[97] For example, Sam Zell acquired the Tribune Co. in a leveraged $8.2 billion deal in 2007; in 2006, McClatchey Co. bought newspaper chain Knight Ridder for more than $4 billion. See CRS Report R40700, The U.S. Newspaper Industry in Transition, by Suzanne M. Kirchhoff.

[98] See, for example, Testimony of W. Kenneth Ferree, President, The Progress & Freedom Foundation, Hearing on "Competition in the Sports Programming Marketplace," Before the Committee on Energy and Commerce, Subcommittee on Telecommunications and the Internet, United States House of Representatives, March 5, 2008, at p. 3.

[99] While major sports events that can be expected to attract a large audience frequently will generate bids for the programming rights to those events, less popular sports events are more likely to be packaged into television programming by syndicators (independent sports television producers who develop programs they sell directly to local stations, sell to advertisers who in turn pay stations to run them, or barter with local stations, with the station and syndicator sharing advertising revenues) or as time buys (in which show producers pay networks or local stations for broadcast time, sell advertising for their shows, and keep all the advertising revenues generated).

[100] Different households may define "favorable access" differently. For example, one household may favor access to a large amount of the desired sports programming, even at a high price, and another household may favor access to sports programming subject to some price ceiling.

[101] Although programming of major sports events attract large audiences, not even the Super Bowl attracts half of U.S. television households in most years—typically attracting 43-48% of households. NFL conference championship games may attract about 30% of television households, the Olympics 18%. Over the decades, only a handful of blockbuster programs—the I Love Lucy show where Lucy went to the hospital to have her baby, Elvis Presley's first appearance on the Ed Sullivan Show, the final M*A*S*H episode, the Dallas episode that answered "Who shot J.R.?," the finale of the Roots mini-series programs, and a CNN program of the O.J. Simpson murder trial—have attracted half of U.S. households. (See Nielsen Media Research and other data presented in the Wikipedia entry on "Highest

rated network telecasts.") Most programming of major sports events attract at most 20% of U.S. television households.

[102] In March 2009, DirecTV signed a contract with the NFL valued at $4 billion for four years for the exclusive rights to sell the Sunday Ticket package. (See Matthew Futterman, "NFL, DirecTV Extend Pact in $4 Billion Deal," The Wall Street Journal, March 24, 2009, at p. B5.) DirecTV offers the service for a monthly charge of $59.95 for five months of the year. The target audience is football fans who are living outside the local viewing area of their favored team as well as football fanatics who want to view games other than the games of their home team. Some of the NFL Sunday Ticket subscribers live in so-called "orphan counties" that are located in one state, but assigned to the local market (known as a "designated market area") for which the principal city and most or all of the local television stations are in another state. Cable and satellite subscribers in these counties may only receive the retransmitted broadcast signals of the stations in their market area, which may not carry the NFL games of their in-state team. Households in these orphan counties can only see the programming of their in-state team if they subscribe to the NFL Sunday Ticket package.

[103] See, for example, Penelope Patsuris, "A Wider World of TV Sports," Forbes.com, December 12, 2002, available at http://www.forbes.com/2002/12/12/cx_pp_1212sports.html, viewed on 08/24/10, which states that "Football, baseball, basketball and hockey often are moneylosing propositions, but they make up for that to some degree since they're excellent platforms for promoting prime-time shows and also help affiliate TV stations boost profits."

[104] SNL Kagan, Broadband Cable Financial Databook, 2009 Edition, table entitled "Census of Basic Cable TV Services (Mil.)," at pp. 14-15.

[105] SNL Kagan, Economics of Basic Cable Networks, 2009 Edition, table entitled "Network Monthly Affiliate Revenue Per Subscriber, By 2010 Average ($)," at p. 53.

[106] Based on data in SNL Kagan, Economics of Basic Cable Networks, 2009 Edition, at p. 2.

[107] See, for example, Michael Malone, "Sinclair, ABC Agree on License Fees for Affiliates," Broadcasting & Cable, March 26, 2010.

[108] Derek Caney, "News Corp posts wider loss on sports write-down," Reuters, February 12, 2002.

[109] See Jon Weisman, "Olympics become NBC loss-leader," Variety, January 29, 2010.

[110] Ibid.

[111] Ibid.

[112] Interestingly, beginning in 2006, the United States Olympic Committee (USOC) sought to launch its own Olympic cable channel. It unsuccessfully pursued creating a partnership with several major program networks before reaching an agreement with Comcast in 2009. Its intention was to create a cable network, which would debut after the conclusion of the January 2010 Winter Olympics in Vancouver, whose programming would include future and past Olympic events and additional programming and for which multichannel video providers would pay monthly per subscriber fees. (This business model of a major sports entity creating its own cable network is discussed below.) When the USOC-Comcast agreement was announced in 2009 it was immediately criticized by the International Olympic Committee. Subsequently, Comcast announced its proposed merger with NBC-Universal, which has been the holder of the U.S. television rights to the Olympic Games, and that dampened Comcast's interest in the cable venture. In April 2010, USOC and Comcast ended their agreement to launch the Olympic cable channel. See Tripp Mickle and John Ourand, "USOC, Comcast End Agreement to Launch New Olympic Channel," SportsBusiness Journal, April 21, 2010. Although the channel is not going forward at this

time, it does indicate that in the future USOC may seek to place all the Olympic events programming on a cable network, rather than keeping the more popular events on a broadcast network.

113 Mass Media Notes, Communications Daily, September 2, 2010.

114 For example, James Robbins, chief executive of Cox Cable, reportedly stated at a Goldman Sachs investors conference in 2003 that ESPN accounted for 4% of Cox subscribers' viewing, but 18% of Cox programming costs. (See David D. Kirkpatrick and Geraldine Fabrikant, "Sports Fan Is the Prize, or the Victim, in Cable Fight," The New York Times, October 6, 2003, at pp. C1 and C4.

115 This has not been the case for sports leagues that do not have a mass fan base, such as the National Hockey League.

116 See Mark Maske, "The NFL's business conundrum," The Washington Post, September 19, 2010, at p. D1.

117 This is a unique subset of the larger issue of tiered vs. a la carte pricing of cable networks. See CRS Report RL32398, Cable and Satellite Television Network Tiering and "a la Carte" Options for Consumers: Issues for Congress, by Charles B. Goldfarb.

In: Broadcast Television in Transition
Editor: Daniel J. Fetonni
ISBN: 978-1-62100-440-0

Chapter 2

How the Satellite Television Extension and Localism Act (STELA) Updates Copyright and Carriage Rules for the Retransmission of Broadcast Television Signals*

Charles B. Goldfarb

Summary

The Satellite Television Extension and Localism Act of 2010 (STELA), P.L. 111-175, modifies the copyright and carriage rules for satellite and cable retransmission of broadcast television signals. The legislation was needed to reauthorize (through December 31, 2014) certain expiring provisions in the Copyright Act and the Communications Act and to update the language in those acts to reflect the transition from analog to digital transmission of broadcast signals, as well as to address certain public policy issues. Had the expiring provisions not been reauthorized, satellite operators would have lost access to a statutory compulsory copyright license and to statutory relief from retransmission consent requirements. This would have made it difficult, if not impossible, for them to retransmit certain distant broadcast signals to

* This is an edited, reformatted and augmented version of a Congressional Research Service publication, CRS Report for Congress R41274, from www.crs.gov, dated January 10, 2011.

their subscribers, including signals providing otherwise unavailable broadcast network programming.

The Copyright Act and Communications Act distinguish between the retransmission of local signals—the broadcast signals of stations located in the same local market as the subscriber—and distant signals. Statutory provisions block or restrict the retransmission of many distant broadcast signals in order to foster local programming. These provisions typically take the form of defining which households are "served" or "unserved" by local broadcasters, with unserved households eligible to receive distant signals. But there are many grandfather clauses and other exceptions built into the rules that allow households to receive otherwise proscribed distant signals. STELA generally retains, and in some cases expands upon, these grandfathered and exceptional cases.

STELA provides broadcasters two new incentives to use their digital technology to broadcast multiple video streams (to "multicast"). It clarifies that royalty fees are payable to copyright owners of the materials on non-primary digital voice streams as well as primary streams, thus encouraging broadcasters (who often hold some of those copyrights) to expand their multicasting. STELA specifically gives broadcasters the incentive to undertake such multicasting to offer otherwise unprovided network programming in so-called "short markets"—markets that do not have network affiliates for all four major networks. It does this by defining households that can receive the programming of a particular network from the non-primary multicast video stream of a local broadcaster as being served, rather than unserved, with respect to that network, thus prohibiting satellite operators from retransmitting to those households distant signals that carry that network's programming. The local broadcaster can then seek retransmission consent payments from satellite operators. Several other provisions in STELA also are intended to reduce the number of short markets or increase the flow of distant network signals into short markets.

Today, satellite operators are allowed, but not required, to offer subscribers the signals of the broadcast stations in their local market. Until enactment of STELA, the satellite operators chose not to offer this "local-into-local" service in many small markets, preferring to use their satellite capacity to provide additional high definition and other programming to larger, more lucrative markets. The costs associated with providing local-into-local service in small markets may exceed the revenues. STELA provided DISH Network, which had been subject to a permanent court injunction that in effect prohibited it from retransmitting to its subscribers the signals of distant broadcast stations, the opportunity to have that injunction waived if it provided local-intolocal service in all 210 local markets in the United States, which it began doing on June 3, 2010.

STELA does not address the issue of "orphan counties"—counties located in one state that are assigned to a local market, as defined by the Nielsen Media Research designated market areas, for which the principal city and most or all of the local broadcast stations are in another state.

OVERVIEW OF STELA

The Satellite Television Extension and Localism Act of 2010 (STELA), P.L. 111-175,[1] extends, updates, and modifies provisions in the Copyright Act[2] and the Communications Act[3] relating to the retransmission of broadcast television signals by satellite television and cable television providers. Among other things, STELA:

- Reauthorizes through December 31, 2014, expiring provisions that provide satellite carriers access to a simple statutory compulsory copyright license and free satellite carriers from retransmission consent requirements, when retransmitting to their subscribers the signals of certain broadcast stations located outside the subscribers' local markets ("distant signals"). Had these provisions expired, it would have been difficult, if not impossible, for satellite operators to provide to their subscribers broadcast network programming that the subscribers are unable to receive from their local broadcasters.
- Revises provisions in copyright and communications law to take into account the transition from analog to digital transmission of broadcast signals.
- Creates an incentive for broadcasters, who often hold copyrights on of the programming they broadcast, to use their digital capabilities to offer multiple video streams ("multicasting") by requiring satellite operators to pay royalty fees for the programming on the non-primary, as well as primary, video streams.
- Provides local broadcasters in markets that currently do not have network affiliates for all four major networks (so-called "short markets") the incentive to offer the programming of the currently unavailable networks on their non-primary digital video streams. Specifically, STELA defines households that can receive the programming of a particular network from the non-primary multicast video streams of a local broadcaster as being "served" rather than "unserved" with respect to that network, thus prohibiting satellite

operators from retransmitting to those households distant signals that carry that network's programming and allowing the broadcaster to seek retransmission consent payments.

- Frees DISH Network of a permanent court injunction against retransmitting the signals of distant network stations into short markets in exchange for the requirement to make available to its subscribers in *each of the 210 local markets* in the United States the signals of all the full-power broadcast stations in the local market. To meet that requirement, on June 3, 2010, DISH began providing such "local-into-local" service to the 29 local markets it had not been serving.
- Modifies the rules governing which households are eligible to receive distant signals from satellite carriers, generally grandfathering those households that currently receive such signals. These rule changes, which attempt to better reflect the current market and technological environment, may increase the number of households that qualify to receive distant signals.
- Modifies the copyright administrative procedures, reporting requirements, royalty fees, filing fees, and non-compliance penalties for the improper retransmission of broadcast television signals by both satellite carriers and cable operators.
- Changes the statutory licenses applicable to the copyrighted material on the retransmitted signals of "significantly viewed" broadcast stations,[4] low power broadcast stations, and other statutorily exceptional[5] stations.
- Requires satellite operators to make available to their subscribers all the programming of non-commercial television stations that is in high-definition format.
- Requires the Register of Copyrights to submit a report on market-based alternatives to statutory licensing and also requires the Comptroller General to submit a report on changes to carriage requirements currently imposed on multichannel video programming distributors (MVPDs) and to Federal Communications Commission (FCC) regulations that might be required if Congress were to phase-out the current statutory satellite and cable licensing requirements.

STELA does not address the situation in which a county has been assigned to a local market for which the principal city is in another state and the television stations located in that local market primarily address the needs of households in that other state, rather than providing news, sports, and other programming of interest to the county. There had been a number of legislative proposals intended to address this "orphan county" issue, but none was included in STELA. But STELA does require the FCC to submit a report on the in-state broadcast programming available to households that receive the signals of broadcast stations that are considered, by statute and rule, to be local but are located in a different state.

BACKGROUND

Congress has constructed a regulatory framework for the retransmission of broadcast television signals by satellite television operators through a series of laws—the 1988 Satellite Home Viewer Act (SHVA),[6] the Satellite Home Viewer Act of 1994,[7] the 1999 Satellite Home Viewer Improvement Act (SHVIA),[8] the 2004 Satellite Home Viewer Extension and Reauthorization Act (SHVERA),[9] and most recently STELA. These laws have fostered satellite provision of MVPD service and, as satellite has become a viable competitor to cable television, have attempted to make the regulatory regimes for satellite and cable more similar. Today, the regulatory framework for satellite exists alongside an analogous, but in some significant ways different, regulatory framework for cable.[10]

The various provisions in these satellite acts created new sections or modified existing sections in the Copyright Act and the Communications Act of 1934. Under current law, in order to retransmit a broadcaster's signals to its subscribers, a satellite operator or a cable operator, with certain exceptions, must obtain a license from the copyright holders of the content contained in the broadcast for use of that *content* and also must obtain the consent of the broadcaster for retransmission of the broadcast *signal*. The statutory provisions addressing copyright are in the Copyright Act and are administered by the Copyright Office in the Library of Congress; those provisions addressing signal retransmission are in the Communications Act and are administered by the FCC. But in several cases, the provisions in one act are conditioned on meeting conditions prescribed in the other act or meeting rules adopted by the agency that administers the other act.

The satellite and cable regulatory frameworks attempt to balance a number of longstanding, but potentially conflicting, public policy goals—most notably, localism, competitive provision of video services, support for the creative process, and preservation of free over-the-air broadcast television.

They also attempt to balance the interests of the satellite, cable, broadcast, and program content industries. Congress incorporated sunset provisions in SHVERA—and again in STELA—because of its concern that market changes could affect these balances. Indeed, as Congress debated the legislative proposals that were included in, or left out of, STELA, it gave substantial weight to a proposed package of changes in copyright procedures, royalty rates, and other parameters constructed and supported by a wide range of industry players through a process of direct negotiations and compromise.

The statutory provisions distinguish between the retransmission of *local* signals—the broadcast signals of stations located in the same local market (as defined by the 210 designated market areas into which the United States is divided by Nielsen Media Research) as the subscriber—and of *distant* signals. These provisions block or restrict the retransmission of many distant broadcast signals in order to protect local broadcasters from competition from distant signals and to provide them with a stronger negotiating position vis-à-vis the satellite and cable operators. The intent is to foster local programming. But the statutory framework also recognizes that U.S. households benefit from the receipt of certain distant broadcast signals and includes explicit retransmission and copyright rules for these.

The statutory framework for satellite sets the parameters within which industry players must conduct business. It provides answers to four fundamental business questions:

- May—or must—a satellite operator retransmit some or all local broadcast signals?[11]
- May a satellite operator retransmit certain categories of distant (non-local) broadcast signals?
- Is retransmission of those signals contingent on a satellite operator receiving the prior retransmission consent of—and providing compensation to—the broadcaster? and
- Is use of the content on those signals subject to specific copyright license terms?

Satellite operators and broadcasters also must conduct business within the constraints of longstanding industry practice. Broadcast program suppliers—both broadcast networks and owners of non-network, syndicated programming—contractually grant individual broadcast television stations the exclusive broadcast rights to their programming in a geographic area and restrict those broadcast stations from allowing other parties to retransmit the station signals carrying that programming beyond the area of exclusivity. Thus, in some situations where the regulatory framework allows satellite (or cable) operators to retransmit the signals of a distant broadcast station, subject to obtaining the permission of the broadcast station, that station may be—and, in practice, often is—contractually prohibited from granting the MVPD retransmission consent.

Although satellite and cable operators compete directly with one another in most markets, there are significant differences in the regulatory frameworks under which they operate. These differences largely reflect the different origins of the cable and satellite industries—cable beginning as a business with technology focused on serving narrow geographic areas and satellite beginning as a business with technology serving broad geographic areas. To this day, cable network architecture and technology can more efficiently accommodate local programming than can satellite. Some observers have proposed that the retransmission, copyright, and other rules under which these competing multichannel video programming distributors operate should be rationalized to eliminate artificial competitive advantages or disadvantages. For example, the Copyright Office, in a report to Congress required by SHVERA,[12] has proposed that the gross receipts royalty system for cable retransmission of distant broadcast signals in section 111 of the Copyright Act be replaced by a flat fee per subscriber system of the sort for satellite retransmission of distant broadcast signals in section 119 of the Copyright Act. The Copyright Office also has proposed[13] that the provisions defining satellite subscriber eligibility for receiving distant signals in section 119 (the "unserved household" provisions) be replaced by the imposition on satellite operators of the FCC's network non-duplication[14] and syndicated exclusivity rules,[15] which currently are used to limit the retransmission of distant broadcast signals by cable operators. But in the Congressional deliberations leading to passage of STELA, there was little discussion of a major modification of the regulatory framework.

Issues Addressed in STELA

Reauthorization

STELA extends through December 31, 2014, several statutory copyright and communications provisions, required for satellite operators to retransmit distant signals, that would have expired on May 31, 2010. Most significantly:

- Section 119 of the Copyright Act[16] provides satellite operators that retransmit certain "distant" (non-local) broadcast television signals to their subscribers with an efficient, relatively low cost way to license the copyrighted works contained in those broadcast signals—a statutory per subscriber, per signal, per month royalty fee. Had the law expired, it would have been very difficult (and perhaps impossible) for satellite operators to offer the programming of broadcast networks[17] to that subset of subscribers who currently cannot receive that programming from local broadcast stations that are affiliated with those networks.[18] It also would have been difficult for satellite operators to offer their subscribers the signals of distant stations that are not affiliated with broadcast networks, including both "superstations"[19] and other non-network stations.
- In addition, prior to the enactment of STELA, section 119 provided those satellite operators that retransmit to their subscribers the signals of "significantly viewed" stations—stations that are located outside the local market in which the subscriber is located but have been determined to be "significantly viewed" by those households in the local market that do not subscribe to any MVPD provider—a royalty-free license for the copyrighted works contained in those broadcast signals. Had section 119 expired, it would have been very difficult (and perhaps impossible) for satellite operators to offer their subscribers the signals of significantly viewed stations. Under STELA, satellite retransmission of significantly viewed stations has been moved from section 119 to section 122 of the Copyright Act, under which such retransmission is subject to the royalty-free license in section 122.
- Section 325(b)(2)(C) of the Communications Act[20] allows a satellite operator to retransmit the signals of distant network stations, without first obtaining the retransmission consent of those distant stations, to those subscribing households that cannot receive the signals of local broadcast television network affiliates. Had it expired, a satellite operator would have had to negotiate compensation terms with those distant network stations whose signals it retransmitted to those "unserved" subscribers.
- Section 325(b)(3)(C)(ii) of the Communications Act[21] prohibits a television broadcast station that provides retransmission consent from engaging in exclusive contracts for carriage or failing to negotiate in good faith. Section 325(b)(3)(iii)[22] prohibits an MVPD from failing to

> negotiate in good faith for retransmission consent. Had these provisions expired, a broadcaster or an MVPD could have chosen to employ a "take it or leave it" strategy rather than to negotiate retransmission consent terms in good faith, increasing the risk of an impasse that results in subscribers losing access to the broadcast station's programming.

STELA includes a provision making the effective date of the act February 27, 2010, in order to protect satellite operators from potential lawsuits for copyright infringement for the brief period of time when the old authorization had expired and Congress had not yet enacted new authorization. At that time, Congress had encouraged the satellite operators not to discontinue retransmission of the distant signals in order to allow satellite subscribers to continue to receive those signals.

Revising Existing Rules that Are Based on Analog Technology

A number of statutory provisions, and many FCC and Copyright Office rules adopted to implement statutory provisions, have been based on the transmission of analog broadcast signals, but during 2009 the transition to digital broadcast signals was largely achieved. As a result, statutes and rules that explicitly referred to analog technology were no longer effective in attaining the objectives for which they were enacted. Thus, Marybeth Peters, Register of Copyrights, proposed five modifications to section 111 of the Copyright Act and four modifications to section 119 of the Copyright Act "to accommodate the conversion from analog to digital broadcasting."[23] Analogous changes were proposed for the Communications Act.

STELA includes specific changes to language in the Copyright Act and to the Communications Act intended to make them consistent with a digital environment. It also includes provisions directing the FCC to develop a predictive model for the reception of digital signals within six months of enactment in order to determine which households are "unserved" and therefore eligible to receive digital network signals. On November 23, 2010, the FCC adopted rules creating measurement standards for digital television signals and establishing a predictive model.[24] STELA also includes a provision that provides guidance for the period before the new predictive model has been implemented

STELA modifies the methodology used to determine whether a household is served to reflect the current market and technological environment, including the transition from analog to digital transmission. It is possible that some of the methodological changes may increase the number of households eligible to receive distant network signals.[25] For example, most households now receive their broadcast signals from their cable or satellite service rather than over-the-air and therefore do not use a rooftop antenna. The old definition of unserved household referred to the inability to receive a signal of a specified intensity using a rooftop antenna; STELA changes the definition to refer to *any* antenna. Since indoor antennas, such as "rabbit-ear" antennas, tend to be less effective than rooftop antennas, this may increase the number of households that qualify as unserved.

Fostering Digital Multicasting, Especially Multicasting to Provide Network Programming in those Markets that Lack a Network Affiliate ("Short Markets")

Although each of the four major broadcast television networks (ABC, CBS, FOX, and NBC) has a local station affiliate in most U.S. markets, 58 of the 210 markets do not have the full complement of four network affiliates.[26] In these short markets, subscribers have been defined as being "unserved" with respect to the missing network and satellite operators have been allowed to retransmit to their subscribers the signals of up to two distant stations that are affiliated with that missing network.[27]

With the transition from analog to digital technology, however, broadcast stations are able to broadcast multiple video streams. Some local television stations in short markets are affiliated with a national network and broadcast that network's programming on their primary video stream, but also have reached agreements with a second national network that lacks an affiliate in the local market to carry the network programming of that second network on a non-primary video stream. This multicasting allows households in the local market to receive the network programming of that second network, although it is unlikely that the local station provides any original local programming on that secondary video stream.

Under STELA, if a local television station broadcasts a non-primary video stream that provides the programming of a national network and was carried by a satellite operator on March 31, 2010, and if the local station continues to carry that network's programming on that video stream, then as of October 1,

2010, that video stream is considered a "qualified multicast video" and households in that local market will be considered served with respect to the broadcast network whose programming is carried on that video stream. Thus, after October 1, 2010, a satellite operator cannot use the statutory distant signal copyright license to retransmit to households in that local market the signal of a distant broadcast station affiliated with that streamed network. Presumably, the satellite operator would have to obtain retransmission consent from the local broadcaster (which probably would entail making a payment to the broadcaster) to retransmit the programming as part of its local-into-local service.

As of January 1, 2011, all non-primary video streams of national network programming offered by a local television station are considered qualified multicast video and households in the local market are considered served with respect to the broadcast network whose programming was carried on those video streams.[28] As a result of this change in treatment of network programming broadcast over non-primary video streams, satellite operators are allowed to retransmit the programming as part of their local-into-local service offering (if they successfully negotiated a retransmission consent agreement with the broadcaster), but are no longer able to retransmit that network programming using a distant broadcast signal.

STELA allows a satellite subscriber who was lawfully receiving the distant signal of a network station on the day before enactment of the new legislation to receive both that distant signal and the local signal of a network station affiliated with the same network until the subscriber chooses to no longer receive the distant signal from its satellite operator. Thus, if in a short market a local broadcaster began to multicast on a non-primary video stream the programming of the network for which there has been no local affiliate, and the satellite operator chose to retransmit that non-primary video stream, a subscriber who has been receiving the distant network signal could continue to receive that distant signal as well as the local network signal, as long as the subscriber did not discontinue its subscription for that distant signal. A household in that short market would not be allowed to receive a distant network signal, however, if it received from the satellite operator the programming of that same network from the non-primary video stream of a local broadcaster but was not a subscriber lawfully receiving the distant signal on the day before enactment of the new legislation.

Another provision in STELA fosters multicasting in all markets, not just short markets. It encourages broadcasters to offer programming over multiple digital video streams—both their primary stream and non-primary streams—

by clarifying that satellite operators must pay copyright royalty fees for the retransmission of the programming on broadcasters' non-primary as well as primary video streams. Since broadcasters often hold some copyrights for the programming they broadcast, such payments increase their incentive to multicast.

Providing an Incentive for DISH Network to Offer Local-into-Local Service in All Designated Market Areas: Allowing DISH to Use a Statutory License to Retransmit Distant Network Signals into Short Markets

Satellite operators are allowed, but not required, to offer subscribers the signals of all the broadcast stations in their local market. If a satellite operator chooses to retransmit the signal of a local broadcast station and to take advantage of a royalty-free statutory copyright license for the content carried on that signal, it must retransmit the primary signals of all the full power stations in that local market, subject to obtaining local station permission. The satellite operators have chosen not to offer this "local-into-local" service in many small markets, preferring to use their satellite capacity to provide additional high definition and other programming to larger, more lucrative markets than to use the capacity to serve very small numbers of customers. In some cases, those small markets may not generate enough revenues to cover the costs of providing local-into-local service.[29] As a result, approximately 3% of all U.S. households do not have access to any local broadcast signals if they subscribe to satellite video service, unless they can receive those signals directly over-the-air.[30]

Early in the 111th Congress, Representative Stupak had introduced a bill, the Satellite Consumers' Right to Local Channels Act, which, in effect, would have required satellite operators to offer local-into-local service in all markets; if a satellite operator wished to use the royalty-free statutory copyright license to rebroadcast the content on a broadcast signal in *any* local market, it would have had to provide local-into-local service in *every* market. But during markup of the House Energy and Commerce Committee bill, Representative Stupak agreed to withdraw his bill (which he had introduced in the form of an amendment), when DISH Network indicated that it would voluntarily provide local-into-local service in all 210 markets within two years in exchange for statutory relief from a current court injunction prohibiting it from providing its subscribers distant signals using the section 119 copyright license.[31] That *quid pro quo* was incorporated into STELA.

As a result of repeated violations of section 119 of the Copyright Act, DISH Network had been subject to a permanent injunction, imposed by the U.S. Court of Appeals for the 11th Circuit,[32] barring it from using the section 119 statutory license for the copyrighted materials when retransmitting distant signals to its subscribers; it therefore had to employ an arms-length agreement with National Programming Service for that entity to deliver distant signals to its subscribers. Under STELA, the injunction is partially waived if DISH Network provides localinto-local service in all 210 local markets in the United States. Specifically, DISH is allowed to use a section 119 license for the copyrighted materials when retransmitting to its subscribers in a "short market" the signals of a distant network broadcast station affiliated with a network for which no local broadcaster is providing the network programming over its primary video stream.

Because of DISH's long history of illegally retransmitting distant signals, STELA incorporates a number of safeguards. DISH must demonstrate that it is offering local-into-local service in all 210 local markets in the United States (referred to as designated market areas or DMAs) in order to be deemed qualified by the court for a temporary waiver of the injunction. The Court must select a special master who would make an initial examination and provide on-going monitoring to assure that DISH is serving all 210 DMAs (and if not, make a determination that it is nonetheless acting reasonably and in good faith) and is in compliance with the royalty payment and household eligibility requirements of the license. The initial waiver of the injunction would be temporary, but could be extended for good cause; if DISH lost recognition as a qualified carrier it could not seek to be re-qualified. Also, the Comptroller General is instructed to monitor the degree to which DISH is complying with the special master's examination. DISH would have the burden of proof that it is providing local-into-local service with a good quality satellite signal to at least 90% of the households in each DMA. It would be subject to penalties of between $250,000 and $5 million for failure to provide service, with exceptions for nonwillful violations.

On June 3, 2010, DISH introduced local-into-local service in the 29 DMAs it had not been serving. These markets are: Alpena, MI; Biloxi, MS; Binghamton, NY; Bluefield, WV; Bowling Green, KY; Columbus, GA; Elmira, NY; Eureka, CA; Glendive, MT; Greenwood, MS; Harrisonburg, VA; Hattiesburg, MS; Jackson, TN; Jonesboro, AR; Lafayette, IN; Lake Charles, LA; Mankato, MN; North Platte, NE; Ottumwa, IA; Parkersburg, WV;

Presque Isle, ME; Salisbury, MD; Springfield, MA; St. Joseph, MO; Utica, NY; Victoria, TX; Watertown, NY; Wheeling, WV, and Zanesville, OH. On June 30, 2010, DISH filed with the FCC an application for certification as a qualified carrier pursuant to section 206 of STELA. On September 1, 2010, the FCC adopted an order granting that certification.[33]

STELA also requires each satellite carrier to submit a semi-annual report to the FCC setting forth (1) each market in which it offers local-into-local service; (2) detailed information regarding the use of satellite capacity for the provision of local-into-local service; (3) each local market in which it has commenced offering local-into-local service in the six-month period covered by the report; and (4) each local market in which it has ceased to offer local-into-local service in the six-month period. The FCC is required to submit to the Senate Commerce, Science, and Transportation Committee and the House Energy and Commerce Committee within a year a report containing the findings, conclusions, and recommendations of a study of (1) incentives that would induce a satellite carrier to offer local-into-local service in markets the carrier has not been serving; and (2) the economics and satellite capacity conditions affecting delivery of local signals by satellite carriers to these markets. On November 23, 2010, the FCC's Media Bureau released a public notice seeking comment and data for use in preparation of that report.[34]

Reducing the Number of Short Markets by Eliminating the "Grade B Bleed" Problem

Prior to enactment of STELA, in areas where a network-affiliated broadcast station was located near the DMA boundary, so that its signal extended into a portion of a neighboring DMA that did not have a local station affiliated with the same network, households in that neighboring market who could receive that signal at a Grade B level were not considered to be "unserved" for that network. A satellite operator could neither offer that overlapping signal to those households as part of local-into-local service (since it was a distant signal) nor provide to those households the signal of a distant station affiliated with the same network, because those households were not considered unserved. The satellite operators sought to eliminate this so-called "Grade B bleed" problem by modifying the test for a subscriber being unserved to apply only to the strength of the signal from an in-market station or by defining unserved in terms of whether the viewer can get local service from the satellite spot beam, rather than in terms of over-the-air reception.[35]

STELA eliminates the problem by defining as "unserved" those households that do not receive the network programming from an over-the-air signal that originates in the local market, that is the signal of their *local network affiliate.*

Household Eligibility to Receive Distant Signals: Grandfathered Subscribers, Other Subscribers, and Households that Are Not Subscribers When Legislation Is Enacted ("Future Applicability")

The primary mechanism for limiting satellite retransmission of distant network signals has been to restrict such retransmission to "unserved" households that cannot receive the programming of a particular network because either (1) the satellite operator is not offering local-into-local service in that market and the households cannot receive a signal of a threshold quality level over-the-air from the local network affiliate, or (2) there is no local affiliate offering the programming of that network. But both the Copyright Act and the Communications Act include certain grandfathered exceptions to those eligibility restrictions; as a result, many households that are able to receive a network signal from a local broadcast station are allowed to continue to receive the distant signal of a broadcast station affiliated with the same network. STELA retains most of these grandfathered exceptions and in some ways expands on them.

Section 339 of the Communications Act sets the rules for carriage of distant television station signals by satellite operators. Section 339(a)(2) addresses the replacement of distant signals with local signals, enumerating four different sets of rules: for grandfathered subscribers to analog distant signals, for other subscribers to analog distant signals, for households that are not subscribers at the time the legislation is enacted (future applicability), and for subscribers to distant digital signals. STELA:

- Retains section 339(a)(2)(A), the grandfathering provision that allows certain households that historically had been receiving distant network signals illegally (and therefore otherwise would not have qualified to receive those distant signals) to continue to receive those signals. The language has been updated only to reflect the date of enactment of the new legislation and to eliminate reference to analog technology. All these households continue to be grandfathered to receive distant network signals despite being able to receive the signals of local stations with the same network affiliation.

- Eliminates references to analog signals from section 339(a)(2)(B), but otherwise the two-part provision is retained. Under the first part, if a household's satellite operator had made a local network station available on January 1, 2005, as part of local-into-local service, the operator would nonetheless be allowed to provide to that household a distant signal of a station affiliated with the same network if the operator had submitted to the television network no later than March 1, 2005, a list of households receiving that distant signal that included that household. This continues the grandfathering of households that had been legally receiving a distant network signal and were allowed to continue to receive that signal when they also had access to the signal of a local broadcast station affiliated with the same network. Under the second part, if the satellite operator had not made available a local network station on January 1, 2005, as part of local-into-local service, the operator would be allowed to offer the household the distant network signal only if (a) the household seeks to subscribe to the distant signal before the date on which the operator begins to offer local-into-local service, and (b) the operator submits to each television network within 60 days of commencing such service the households subscribing to the distant signal. Thus, a household that had legally sought to receive a distant network signal is allowed to continue to receive that signal after the signal of a local broadcast station affiliated with the same network is available.
- Allows a subscriber who is lawfully receiving the distant signal of a network station from a satellite operator on the day before enactment of STELA to receive both the distant signal and the local signal of the same network until the subscriber chooses to no longer receive the distant signal from the satellite operator (whether or not the subscriber elects to subscribe to local-into-local service). Thus, all the households legally receiving distant network signals under section 339(a)(2)(B) at the date of enactment of STELA continue to be allowed to receive those distant signals.
- Prohibits a satellite operator from providing a distant network signal to a person who (1) (a) is not a subscriber legally receiving that distant signal on the date STELA is enacted, and (b) at the time the person seeks to receive the distant signal, resides in a local market where the satellite operator offers local-into-local service that includes a local

station affiliated to the same network and the person can receive that local-into-local service, *or* (2) (a) is a subscriber legally receiving a distant signal on or after the date STELA is enacted, and (b) subsequent to such subscription the satellite carrier makes available to that subscriber the signal of a local network station affiliated with the same network as the distant signal (and the retransmission of such signal by the carrier can reach the subscriber), unless the person subscribes to the signal of the local network station within 60 days after the signal is made available. The latter is intended to support local stations by requiring the subscriber to obtain local-intolocal service in order to continue to receive the distant network signal.

- Defines a subscriber as eligible to receive a distant signal of a network station affiliated with the same network as a local station if, with respect to a local network station: (1) the subscriber's household is not predicted by the model specified in the act to receive the threshold signal intensity; (2) the household is determined, based on a test conducted in accordance with the current model or any successor regulation, not to be able receive the signal of the local station with an intensity that exceeds the standard; or (3) the subscriber is in an unserved household as determined by the definition of an unserved household in section119(d)(10)(A) of the Copyright Act. The third criterion appears to allow a household that does not meet the signal intensity test for analog service for the signal of a local network station to be grandfathered for the receipt of a distant network signal carrying the same network, even if that household could receive the digital signal of the local network station.

Provisions in section 119 of the Copyright Act define "unserved households" and set the copyright rules that apply to the secondary transmission of distant signals to those unserved households. STELA modifies some of those provisions.

- If a local station is multicasting and offers a second network's programming on one of its non-primary video streams, but a household using an antenna cannot receive that non-primary video stream at the signal intensity specified in FCC rules, then the household is deemed unserved with respect to the network whose programming is being broadcast on that non-primary stream. This took effect on October 1, 2010, for multicast streams that existed on

March 31, 2010, and on January 1, 2011, for all other multicast streams.

- References to analog signals are eliminated, but otherwise all the rules covering grandfathered households receiving distant signals currently in section 119(a)(4)(A) are retained.
- For a subscriber (other than a grandfathered household) who, on the day before enactment of STELA, was lawfully receiving a satellite retransmission of a distant network signal under a statutory license, the statutory license shall apply for the retransmission of that distant signal. Further, the subscriber's household shall continue to be considered an unserved household with respect to that network until the subscriber elects to stop receiving that distant signal, whether or not the subscriber has access to the signal of a local network station affiliated with the same network through local-into-local service and whether or not the subscriber elects to subscribe to that local-into-local service. This, in effect, creates a new group of grandfathered households.
- The statutory distant signal copyright license in section 119 of the Copyright Act does not apply to the satellite retransmission of a distant network signal to a person who is not a subscriber lawfully receiving that distant network signal at the date of enactment of STELA if, when that person subsequently seeks to subscribe to a satellite carrier for that distant signal, that person can obtain that network's programming from a local station affiliated with the same network through local-into-local service.
- The statutory distant signal copyright license in section 119 of the Copyright Act applies to the satellite retransmission of a distant network signal to a person who is a subscriber lawfully receiving that distant network signal on or after the date of enactment of STELA, and the subscriber's household continues to be considered to be an unserved household with respect to that network, until such time as the subscriber elects to terminate such retransmission, but only if the person subscribes to retransmission of a local network station affiliated with the same network (that is, subscribes to local-into-local service) within 60 days of the satellite carrier making local-into-local service available to the subscriber. Thus, a household can be grandfathered for the distant network service only if it subscribes to local-into-local service within 60 days of that service becoming available.

Modified Copyright Treatment of the Satellite Retransmission of Low Power Television Station Signals

Low power television service was created by the FCC in the 1980s to serve small communities (rural or urban) with low cost, limited geographic range facilities that used available spectrum between full power stations. It is a "secondary service" that is not guaranteed protection from interference or displacement by full service stations Low power stations that produced at least two hours per week of local programming, maintained a production studio within their Grade B contour, and complied with many of the requirements placed on full service stations were given a one-time opportunity to obtain "Class A" status that gave them primary status, that is, protected their channel from interference or displacement.

Historically, satellite retransmission of low power television signals was covered by the statutory distant signal copyright license in section 119 of the Copyright Act. Satellite operators were allowed to retransmit the signals of low power stations to subscribers within certain geographic limitations—to subscribers within 20 miles of the station transmitter for network-affiliated stations located in the 50 largest markets, within 35 miles of the station transmitter for network-affiliated stations located in other markets, and within the same designated market area as nonnetwork-affiliated stations.[36] Satellite operators had no copyright royalty obligation for retransmission of the low power station content within those same mileage limits; beyond those limits, satellite operators were subject to the statutory copyright license fees for distant signals outlined in section 119 of the Copyright Act.[37]

Under STELA, if satellite operators seek to use a statutory license for the copyrighted material on the low power television stations whose signals they retransmit, they must use the royalty-free statutory local signal license in section 122, rather than the section 119 license. STELA expands the geographic area covered by the royalty-free statutory license to the entire DMA in which the low power station is located.

STELA also explicitly states that a satellite carrier that retransmits the signal of a low power station under a statutory license is not required to make any other secondary retransmissions. Thus retransmission of a low power station does not trigger the requirement to offer local-intolocal service or to retransmit any other low power stations. No local low power station can demand carriage by the satellite operator serving its market area, even if that satellite operator is providing local-into-local service.

Since low power television stations do not have a deadline for their transition from analog to digital transmission, the old, analog-based FCC rules for determining whether a household is eligible to receive distant signals will apply with respect to low power television until the station is licensed to broadcast a digital signal.

The statutory local signal copyright license does not apply to satellite retransmission of repeaters or translators.

Satellite Carriage of Noncommercial Educational Television Stations

By statute, providers of direct broadcast satellite service (DirecTV and DISH Network) must reserve between 4% and 7% of their channel capacity exclusively for noncommercial programming of an educational or informational nature.[38] With the digital transition, broadcasters now are able to broadcast high definition signals and multiple digital programming streams over their licensed spectrum, and the public television stations are seeking to expand satellite carriage of their high definition and multicast signals.

The public broadcasters have reached retransmission consent agreements with DirecTV, the cable industry (through both the National Cable and Telecommunications Association representing large cable operators and the American Cable Association representing small cable operators), and Verizon for the retransmission of most of their high definition and multicast video streams. The agreement with DirecTV incorporated "creative solutions that recognized [DirecTV's] capacity limitations; ultimately ensuring that subscribers have access to the myriad of content and services provided by the local stations while accommodating their capacity concerns."[39] The public broadcasters have not yet achieved retransmission agreement with DISH Network, but negotiations are continuing.

STELA modifies section 338(a) of the Communications Act, which addresses the carriage of local television signals by satellite carriers, to require any satellite carrier that has not yet negotiated a carriage contract covering at least 30 noncommercial educational television stations by July 27, 2010, (1) to provide subscribers, by the end of 2011, the high definition signals of qualified noncommercial educational television stations in all the local markets in which the carrier currently offers local television broadcasts in high definition (and by the end of 2010 to half of those markets), and (2) when initiating the provision of high definition local broadcast television in a market, to include the high definition signals of all qualified local noncommercial educational television stations. The only satellite carrier that did not have such a carriage contract was DISH Network.

In July 2010, DISH filed suit in the U.S. District Court for Nevada, seeking temporary injunctive relief from FCC enforcement of that provision, claiming the provision infringed on its First Amendment right "to make the editorial judgment whether to carry local PBS stations in HD" and was confiscatory of its property.[40] U.S. District Judge James Mahan declined DISH's request for an injunction.[41] DISH did not reach an agreement with Association of Public Television Stations, which represents all public television stations, but it averted the carriage mandate in STELA by reaching an independent HD carriage agreement with 30 noncommercial stations by July 27, 2010.[42]

At an October 7, 2009, hearing of the Senate Subcommittee on Communications, Technology, and the Internet, public broadcasters identified another problem for which they sought a legislative solution. Most states have developed state public television networks intended to serve the entire state, but in 16 states those networks do not have public stations transmitting signals in each DMA in the state; under current law, satellite carriers are not allowed to use a royalty-free statutory copyright license to retransmit the signals of the in-state, but out-of-market public broadcasting stations to their subscribers in those DMAs.[43] STELA modifies the provisions for the royalty-free statutory copyright license in section 122 of the Communications Act to allow, where there is a public educational network of three or more noncommercial educational broadcast stations in a state, a satellite operator to use the royalty-free license to retransmit the programming on those stations' signals to subscribers in any county in the state whose households are otherwise ineligible to receive retransmissions of those signals.

Satellite Carriage of State Public Affairs Networks

Cable franchise authorities are allowed, by law, to require cable operators to set aside some of their capacity for the carriage of public, educational, and governmental (PEG) programming. This programming is not broadcast to the public; it is sent directly to the cable system's head-end for retransmission. Satellite operators are not required to offer PEG programming, though they have the obligation to allocate between 4 and 7 percent of their channel capacity exclusively to noncommercial programming of an educational or informational nature. In order to foster PEG programming on cable systems, a number of states have created state public affairs networks that produce non-broadcast programming of state-wide interest. Although this programming is available to satellite operators, those operators are not widely offering it to subscribers.

STELA includes a provision intended to encourage satellite operators to carry these state public affairs networks. Under the provision, a satellite carrier that provides the retransmission of the state public affairs networks of at least 15 different states, under reasonable prices, terms, and conditions, and does not delete any of the noncommercial educational or informational programming on those networks, would only have to allocate 3.5% of its channel capacity to the retransmission of educational or informational noncommercial programming, rather than 4%. This provision might encourage satellite operators to offer state public affairs networks to subscribers in orphan counties or in short markets.

The Retransmission of In-State, but Non-Local, Broadcast Signals into Counties Assigned to Local Markets in Other States ("Orphan Counties")

The current regulatory frameworks for both satellite and cable distinguish between the retransmission of local and distant signals and require that local markets be defined by the DMAs constructed and published by Nielsen Media Research.[44] The viewing patterns that underlie these Nielsen markets are primarily the result of the physical locations of the various broadcast television stations and the reach of their signals. (They also reflect the boundaries of the exclusive broadcast territories that each of the three original television broadcast networks—ABC, CBS, and NBC—had incorporated into their contracts with their local affiliate stations decades ago.) DMAs do not take into account state boundaries. As a result, under current statutes and rules, a number of counties are assigned to local markets for which the principal city (from which all or most of the local television signals originate) is outside their state.[45] Satellite subscribers (and many cable subscribers) in these "orphan counties" may not be receiving signals from in-state broadcast stations and as a result may not be receiving news, sports, and public affairs programming of interest in their state.

Many residents of orphan counties have proposed that the statutory framework be modified to remove prohibitions or impediments on satellite operators retransmitting to their subscribers in these counties the signals of broadcast stations in in-state, but non-local, markets. (SHVERA selectively removed these impediments through four "exceptions" that allow satellite operators to retransmit to their subscribers in particular orphan counties in New Hampshire, Vermont, Oregon, and Mississippi—but not in other locations—the signals of in-state but out-of-market broadcast stations.[46]) Broadcasters, however, have voiced concern that allowing such retransmission could undermine their financial viability by reducing their audience share and thus reducing their advertising revenues. They also assert such retransmission would weaken the local broadcasters' negotiating position with the satellite and cable operators, who could turn to the programming of an in-state but out-of-market affiliate of a particular network if they failed to reach retransmission consent with the local affiliate of that network. Broadcasters claim this would harm their ability to provide quality local programming, which is expensive to produce.[47]

A number of bills had been introduced in the 111th Congress that directly addressed this issue (either generically or for specific states or geographic areas) by allowing satellite operators to retransmit to subscribers in orphan counties the signals of certain in-state, but non-local broadcast stations.[48] But STELA (reflecting each of the four bills that had been reported out of the House Energy and Commerce, House Judiciary, Senate Commerce, Science, and Transportation, and Senate Judiciary committees, leading to STELA) does not include any provisions that would address this issue directly. During the markup of the Senate Judiciary Committee bill, reportedly several Senators prepared amendments that would have narrowly addressed the orphan county issue in their states, but then agreed to withdraw their amendments when other Senators voiced concern that the provisions would delay passage of the legislation because of unresolved issues among broadcasters and satellite operators. At the markup, reportedly there was discussion of imposing a deadline on the industry to reach a negotiated solution, such as a proposal by Senator Coburn that, if there were no industry agreement by the time the legislation reaches the Senate floor, a trigger provision would be inserted in the bill that would impose a statutory solution for the orphan counties issue if no negotiated compromise is reached after two years.[49] But STELA does not include a trigger provision.

STELA includes a provision instructing the FCC to prepare within one year a report containing analysis of (1) the number of households in each state that receive local broadcast signals from stations of license located in a different state; (2) the extent to which consumers have access to instate broadcast programming; and (3) whether there are alternatives to the use of DMAs to define local markets that would provide more consumers with in-state broadcast programming. On November 23, 2010, the FCC's Media Bureau released a public notice seeking comment and data for use in preparation of the report.[50]

In addition, a savings clause in STELA—stating that nothing in the legislation, in the Communications Act, or in any FCC regulation shall limit the ability of a satellite operator to retransmit a performance or display of a work pursuant to an authorization granted by the copyright owner—is intended to clarify that a satellite operator always has the opportunity to negotiate a copyright license outside the section 119 statutory license. This clarification is not likely to result in the satellite retransmission into orphan counties of the sports and network programming on in-state, but out-of-market stations, but could encourage the retransmission of those stations' locally produced news programming.

Changing the Statutory Copyright License Applied to the Content on the Signals of Significantly Viewed and "Exception" Broadcast Stations

The statutory framework for the retransmission of broadcast television signals has been based on a distinction between local and distant signals. The signals of significantly viewed stations and the signals of in-state, out-of-market stations in the four states that satellite operators were allowed to import into orphan counties under the exceptions in SHVERA, originate outside the market into which they are imported; in that regard, they are distant signals and they have been subject to the section 119 distant signal statutory copyright license. But since significantly viewed stations and the "exception" stations can be presumed to be providing programming of local or state-wide interest to counties in particular local markets, arguably that content could be viewed as local to the counties into which they are imported and should be treated accordingly. STELA modifies the Copyright Act to treat those signals as local, moving the relevant provisions from section 119 to section 122.

STELA changes language in the heading of section 122 from "secondary transmission by satellite carriers within local markets" to "secondary transmission of local television programming by satellite." It makes satellite retransmission of both significantly viewed stations and the exception stations subject to the local signal statutory copyright license in section 122 rather than the distant signal statutory license in section 119, but requires the satellite operator to continue to pay the statutory copyright license fees under section 119 for the retransmission of the exception stations. Since significantly viewed stations already are subject to the royalty-free license in section 122, effectively there is no change in copyright treatment for the content on the signals of significantly viewed stations. But the statutory change allows DISH Network, which currently is under a court injunction prohibiting it from using the section 119 statutory copyright license to retransmit the content of broadcast signals, to use the section 122 statutory copyright license to do so.

Although STELA changes the statutory license required for satellite retransmission of the signals of significantly viewed and exception stations, it does not affect the retransmission consent requirement or the exemption from the FCC's network non-duplication and syndicated exclusivity rules, as they currently apply to significantly viewed and exception stations. It does, however, include a provision stating that the satellite operator would not be required to carry the significantly viewed stations or exception stations if it offered local-into-local service.

Allowable Signal Formats for the Retransmission of Significantly Viewed Stations

The satellite operators have complained that although both cable and satellite operators may offer significantly viewed stations, only satellite operators have been subject to an "equivalent bandwidth" provision that, as interpreted by the FCC, required the satellite operator to carry the signals of a significantly viewed station that is affiliated to the same network as a local station in the same format as that local station every moment of the day. Thus, for example, if the local station were not transmitting its programming in high definition format, the satellite operator would not be allowed to retransmit into the market the signals of the significantly viewed station in high definition format. According to satellite operators, this was infeasible.

STELA clarifies that a significantly viewed signal may only be provided in high definition format if the satellite carrier is passing through all of the high definition programming of the corresponding local station in high definition format as well; if the local station is not providing programming in high definition format, then the satellite operator is not restricted from providing the significantly viewed station's signal in high definition format.

Studying What the Impact Would Be If the Statutory Licensing System for Satellite and Cable Retransmission of Distant Broadcast Signals Were Eliminated

The United States Copyright Office has proposed that Congress abolish sections 111 and 119 of the Copyright Law, arguing that the statutory licensing systems created by these provisions result in lower payments to copyright holders than would be made if compensation were left to market negotiations.[51] According to the Copyright Office, the cable and satellite industries no longer are nascent entities in need of government subsidies, have substantial market power, and are able to negotiate private agreements with copyright owners for programming carried on distant broadcast signals.

One possible way to transition from the current licensing system would be to enact a statutory "trigger" mechanism, by which once a broadcast station successfully demonstrated that it had obtained the rights to negotiate for all the holders of copyrighted materials on its programming, so that a satellite carrier did not have to negotiate with multiple copyright holders, the statutory license for that station would sunset and the satellite operator would have to undertake private negotiations. This is strongly opposed by satellite operators, who question how voluntary licensing arrangements and sublicensing would work in practice.[52] Other parties argue that the current licensing systems are efficient and that the purpose of copyright law is to balance the potentially conflicting goals of fostering the dissemination of copyrighted material and allowing the copyright holder to be compensated by giving the copyright holder a *limited* monopoly over its material; they oppose a rule that allows the copyright holder to fully exploit its monopoly power to receive whatever the market would bear.[53]

STELA instructs the Copyright Office, after consultation with the FCC, to submit to the House and Senate Judiciary Committees, within one year, a report containing proposed mechanisms, methods, and recommendations on how to implement a phase-out of the current statutory license requirements in sections 111, 119, and 122 of the Copyright Act, including recommendations for legislative or administrative actions.

STELA also instructs the Comptroller General to prepare and submit a report within 12 months that analyzes and evaluates the changes to the cable and satellite carriage requirements in the Communications Act and in FCC rules that would be required if Congress implemented a phaseout of the current section 111, 119, and 122 statutory licensing requirements in the Copyright Act. It instructs the Comptroller General to consider the impact of such a phase-out on consumer prices and access to programming and to include recommendations for legislative or administrative actions.

Providing Digital Service on a Single Dish

Under section 338(g) of the Communications Act, satellite operators have been required to provide to their subscribers the analog signals of all broadcast stations on a single roof-top dish. Operators have been allowed to use a second dish for the provision of digital signals, but there was no requirement that all digital signals be provided on the same dish. STELA modifies section 338(g) to require a satellite operator, if it offers local-into-local service in a market, to provide to a subscriber the digital signals of all the local broadcast stations on a single dish.

Modification of the Methodology for Setting Copyright Royalty Rates and of Copyright Administrative Procedures and Requirements

STELA modifies the methodology for setting copyright royalty rates as well as copyright administrative procedures and requirements. Among these changes, STELA:

- requires satellite operators whose retransmissions of distant broadcast signals are subject to the section 119 statutory license to pay a filing fee, to be determined by the Register of Copyrights, to help recoup the administrative costs of distributing royalty fees;
- modifies the section 119 statutory royalty fee payable to copyright owners to take into account the non-primary streams of multicasting broadcasters;
- instructs the Register of Copyrights to issue regulations to permit interested parties to verify and audit the statements of account and royalty fees submitted by satellite carriers and cable operators;
- changes the process for adjusting royalty fees. Most significantly, STELA creates a proceeding of the Copyright Royalty Judges, which replaces the previously used compulsory arbitration proceeding, to determine royalty rates;

- instructs the Copyright Royalty Judges, when determining royalty rates in those situations where the parties are not able to reach a negotiated agreement, to establish fees that represent the fair market value of the retransmissions, basing their decision on economic, competitive, and programming information presented by the parties;
- requires the Copyright Royalty Judges to make an annual adjustment to the royalty fee based on the most recent consumer price index for all consumers and for all items;
- increases the statutory maximum damages to be imposed on satellite operators for violating territorial restrictions on the retransmission of distant broadcast signals from $5 to $250 per subscriber per month during which the violation occurred. It also increases the maximum statutory damages for regional or large-scale violations (that do not trigger a permanent injunction) from $250,000 for each 6-month period to $2.5 million for each 3-month period. One half of the statutory damages ordered are to be deposited with the Register of Copyrights and distributed to copyright owners;
- modifies the statutory license for retransmission by cable systems in section 111 of the Copyright Act by increasing the specified percentages of the gross subscriber receipts that cable operators must pay;
- updates the definition of "distant signal equivalent" used to reflect and take into account multicast signals when calculating the cable royalty payment, and sets a schedule for when these changes go into effect based on existing contractual agreements;
- clarifies that the royalty rates specified in sections 256.2(c) and (d) of title 37, Code of Federal Regulations, commonly referred to as the "3.65% rate" and the "syndicated exclusivity surcharge," respectively, do not apply to multicast streams;
- clarifies that when a cable operator retransmits a distant broadcast signal to a service area comprised of multiple communities, in which some communities are permitted to receive that signal and other communities are prohibited to do so, the royalty calculation does not include payment for the households that are not allowed to receive the signal;[54]
- modifies the methodology for determining the maximum and minimum royalty payments for small cable systems; and
- creates filing fees for satellite carriers and cable operators filing statements of account for section 111, 119, and 122 statutory

copyright licenses that are reasonable and that do not exceed one-half of the cost incurred by the Copyright Office for the collection and administration of the statements of account and any royalty fees deposited with the statements.

Severability

STELA includes a "severability" provision stating that if any provision of the new law, amendment made by the new law, or applications of such provision or amendment is held to be unconstitutional, the remainder of the law, amendments, and applications would not be affected. This provision was included because there has been a long history of litigation in this area and is intended to make sure that the entire law would not be overturned if there were a successful legal challenge to one provision.

End Notes

[1] 124 Stat. 1218.

[2] 17 U.S.C. §§ 111, 119, and 122.

[3] 47 U.S.C. §§ 325, 335, 338, 339, 340, and 341.

[4] "Significantly viewed" stations are located outside the local market in which the subscriber is located but have been determined by the Federal Communications Commission to be viewed by a "significant" portion of those households in the local market that do not subscribe to any multichannel video programming distributor (MVPD). The specific threshold viewing level for a significantly viewed station are, for a network affiliate station, a market share of at least 3% of total weekly viewing hours in the market and a net weekly circulation of 25%; for independent stations, 2% of total weekly viewing hours and a net weekly circulation of 5%. The share of viewing hours refers to the total hours that households that do not receive television signals from MVPDs viewed the subject station during the week, expressed as a percentage of the total hours these households viewed all stations during the week. Net weekly circulation refers to the number of households that do not receive television signals from MVPDs that viewed the station for 5 minutes or more during the entire week, expressed as a percentage of the total households that do not receive television signals from MVPDs in the survey area. A satellite operator can retransmit the signals of these significantly viewed stations only with the retransmission consent of the station.

[5] The 2004 Satellite Home Viewer Extension and Reauthorization Act allowed satellite operators to retransmit in-state but non-local broadcast television signals to subscribers located in certain counties in Vermont, New Hampshire, Oregon, and Mississippi that are assigned to local markets (as defined by Nielsen Media Research designated market areas) whose local broadcast stations are located in another state. For convenience, these stations are referred to as statutorily exceptional stations.

[6] P.L. 100-667.

[7] P.L. 103-369.

[8] P.L. 106-113.

[9] P.L. 108-447, passed as Division J of Title IX of the FY2005 Consolidated Appropriations Act.

[10] For a more detailed discussion of the differences in the rules for cable and satellite providers, see CRS Report R40624, *Reauthorizing the Satellite Home Viewing Provisions in the Communications Act and the Copyright Act: Issues for Congress*, by Charles B. Goldfarb, especially at Table 1, "Current Retransmission and Copyright Rules for Satellite and Cable Operators."

[11] This is formally referred to in the statute as "secondary transmission" of the broadcast signals. The initial transmission of the signals by the broadcast station is the "primary transmission."

[12] *Satellite Home Viewer Extension and Reauthorization Act Section 109 Report*, A Report of the Register of Copyrights, June 2008, at pp. ix-xi and 94-180.

[13] *Satellite Home Viewer Extension and Reauthorization Act Section 109 Report*, A Report of the Register of Copyrights, June 2008, at pp. 167-168.

[14] 47 C.F.R. §§ 76.92, 76.93, 76.106, 76.120, and 76.122. Commercial television station licensees that have contracted with a broadcast network for the exclusive distribution rights to that network's programming within a specified geographic area are entitled to block a local cable system from carrying any programming of a more distant television broadcast station that duplicates that network programming. Commercial broadcast stations may assert these non-duplication rights regardless of whether or not the network programming is actually being retransmitted by the local cable system and regardless of when, or if, the network programming is scheduled to be broadcast. This rule applies to cable systems with more than 1,000 subscribers. Generally, the zone of protection for such programming cannot exceed 35 miles for broadcast stations licensed to a community in the FCC's list of top 100 television markets or 55 miles for broadcast stations licensed to communities in smaller television markets. The non-duplication rule does not apply when the cable system community falls, in whole or in part, within the distant station's Grade B signal contour. In addition, a cable operator does not have to delete the network programming of any station that the FCC has previously recognized as "significantly viewed" in the cable community. With respect to satellite operators, the network non-duplication rule applies only to network signals transmitted by superstations, not to network signals transmitted by other distant network affiliates.

[15] 47 C.F.R. §§ 76.101, 76.103, 76.106, 76.120, and 76.123. Cable systems that serve at least 1,000 subscribers may be required, upon proper notification, to provide syndicated protection to broadcasters who have contracted with program suppliers for exclusive exhibition rights to certain programs within specific geographic areas, whether or not the cable system affected is carrying the station requesting this protection. However, no cable system is required to delete a program broadcast by a station that either is significantly viewed in the cable community or places a Grade B or better contour over the community of the cable system. With respect to satellite operators, the syndicated exclusivity rule applies only to syndicated programming transmitted by superstations, not to syndicated programming transmitted by other distant broadcast stations.

[16] 17 U.S.C. §119.

[17] A network is defined as an entity that offers an interconnected program service on a regular basis for 15 or more hours per week to at least 25 affiliated television licensees in 10 or more states. (17 U.S.C. § 119(d)(2)(A) and 47 U.S.C. § 339(d)(2)(A)) In addition to the four major television networks—ABC, CBS, Fox, and NBC—that provide national news and

entertainment programming aimed at a general audience, there are several networks—Univision, Telefutura, and Telemundo—that offer news and entertainment targeted to ethnic communities, as well as smaller networks that provide entertainment or religious programming to their affiliates. Section 119(d)(2)(B) of the Copyright Act defines "network station" to also include noncommercial broadcast stations.

[18] This would include subscribers who are not able to receive network programming because either (1) the satellite operator does not offer the signals of the local broadcast stations and the subscribers are located too far from the transmitter to receive the signals of the local network-affiliated stations over-the-air or (2) there is no network-affiliated station in the local market. The specific household eligibility requirements for receiving distant signals are very complex and include certain grandfathered exceptions, but as a general rule households that can receive the signals of local broadcast television stations either over-the-air or as part of local-into-local satellite service are not eligible to receive distant network signals and would not be affected by the expiration of this provision.

[19] Prior to enactment of STELA, the Copyright Act and the Communications Act both had language referring to "superstations," but that term was defined differently in the two acts, thus creating confusion. The Communications Act identifies a class of "nationally distributed superstations" (47 U.S.C. § 339(d)(2)) that is limited to six stations that were in operation prior to May 1, 1991. These are independent broadcast television stations whose broadcast signals are picked up and redistributed by satellite to local cable television operators and to satellite television operators all across the United States. These nationally distributed superstations in effect function like a cable network rather than a local broadcast television station or a broadcast television network. The nationally distributed superstations are WTBS, Atlanta; WOR and WPIX, New York; WSBK, Boston; WGN, Chicago; KTLA, Los Angeles; and KTVT, Dallas. All of these nationally distributed superstations carry the games of professional sports teams. It has become common in FCC proceedings and discussions to refer to these nationally distributed superstations as simply "superstations." In addition to these independent nationally distributed superstations, there also are many independent television stations that are not nationally distributed superstations. This distinction is important because under section 325(b)(2)(B) of the Communications Act, satellite operators may retransmit the signals of "superstations" without obtaining the consent of the stations if they abide by the FCC's network non-duplication and syndicated exclusivity rules (see footnotes 11 and 12 above), but this exemption from the retransmission consent requirement does apply to the retransmission of the signals of other independent stations. On the other hand, until statutory changes were made in STELA, the Copyright Act had defined "superstation" as "a television station, other than a network station, licensed by the Federal Communications Commission, that is secondarily transmitted by a satellite carrier." (17 U.S.C. § 119(d)(9)) Thus, under the Copyright Act pre-STELA, all independent stations were considered superstations and the copyright provisions applied the same way to all independent stations. Language in STELA eliminated the definitional inconsistency between the acts by replacing the word "superstation" with "non-network station" throughout the Copyright Act.

[20] 47 U.S.C. § 325(b)(2)(C).

[21] 47 U.S.C. § 325(b)(3)(C)(ii).

[22] 47 U.S.C. § 325(b)(3)(C)(iii).

[23] Marybeth Peters, Register of Copyrights, written statement before the House Judiciary Committee, hearing on "Copyright Licensing in a Digital Age: Competition, Compensation and the Need to Update the Cable and Satellite TV Licenses," at Appendix 1, February 25,

2009. The proposed modifications to section 111 include revising section 111, and its terms and conditions, to expressly address the retransmission of digital broadcast signals; amending the definition of "local service area of a primary transmitter" to include references to digital station "noise limited service contours" for purposes of defining the local/distant status of noncommercial educational stations (and certain UHF stations) for statutory royalty purposes; amending the statutory definition of "distant signal equivalent" (DSE) to clarify that the royalty payment is for the retransmission of the copyrighted content without regard to the transmission format; amending the definitions of "primary transmission" and "secondary transmission," as well as the "station" definitions in section 111(f) so they comport to the amended definition of DSE; and clarifying that each multicast stream of a digital television station shall be treated as a separate DSE for section 111 royalty purposes. The proposed modifications to section 119 include replacing the existing Grade B analog standard with the new noise-limited digital signal intensity standard; adopting the Individual Location Longley Rice (ILLR) predictive digital methodology for predicting whether a household can receive an acceptable digital signal from a local digital network station; mandating that the FCC adopt digital signal testing procedures for purposes of determining whether a household is actually unserved by a local digital signal; and deleting various references in section 119 to "analog" unless that reference is to low power television stations that have not yet converted to digital broadcasting.

[24] *In the Matter of Measurement Standards for Digital Television Signals Pursuant to the Satellite Home Viewer Extension and Reauthorization Act of 2004*, ET Docket No. 06-94, Report and Order, adopted November 22, 2010, and released November 23, 2010, and *In the Matter of Establishment of a Model for Predicting Digital Broadcast Television Field Strength Receive at Individual Locations*, ET Docket No. 10-152, Report and Order and Further Notice of Proposed Rulemaking, adopted November 22, 2010, and released November 23, 2010.

[25] See, for example, Lauren Lynch Flick and Scott R. Flick, "Congress Passes Satellite Television Extension and Localism Act of 2010," Pillsbury Winthrop Shaw Pittman LLP Client Alert, May 14, 2010, available at http://www.ilba.org/downloads/~mo~FCC/Congress_Passes_STELA.pdf, viewed on June 2, 2010. Pillsbury is a law firm with many broadcaster clients.

[26] Warren Communications, *Television & Cable Factbook 2010*, Station Volume 2, "Affiliations by Market for TV Stations, as of October 1, 2009," at pp. C-5 – C-8.

[27] 47 U.S.C. § 339. This provision applies to all network stations, but in practice it primarily involves the retransmission of distant signals into short markets that do not have local broadcast stations affiliated with each of the four major national broadcast networks.

[28] There remains a brief transition period, October 1, 2010, to January 1, 2011, during which if a local broadcaster were to begin multicasting another broadcast network signal, the signal would not be deemed a qualified multicast video and a satellite carrier could import into the local market the signal of a broadcaster affiliated with the same network.

[29] Paul Gallant, an analyst with Stanford Washington Research Group, reportedly stated that mandatory provision of local-into-local service in all markets "would impose significant new costs on Dish Network and DirecTV and generate virtually no new revenue" because the markets in question are so small. See Todd Shields, "DirecTV, Dish May Face Requirement for More Local TV (Update1)," Bloomberg.com, February 23, 2009, available at http://www.bloomberg.com/apps/news?pid=newsarchive&sid=ayQ_vo3nJImo, viewed on April 27, 2009.

[30] According to the written testimony of Charles W. Ergen, chairman, president, and chief executive officer of DISH Network Corporation, submitted for the hearing on "Reauthorization of the Satellite Home Viewer Extension and Reauthorization Act," before the Subcommittee on Communications, Technology, and the Internet, Committee on Energy and Commerce, U.S. House of Representatives, February 24, 2009, at p. 2, "DISH provides local service in 178 markets today, reaching 97 percent of households nationwide." According to the written testimony of Bob Gabrielli, senior vice president, broadcasting operations and distribution, DIRECTV, Inc., before the House Judiciary Committee, February 25, 2009, at p. 10, "DIRECTV today offers local television stations by satellite in 150 of the 210 local markets in the United States, serving 95 percent of American households. (Along with DISH Network, we offer local service to 98 percent of American households.)"

[31] See John Eggerton, "DISH: Local Into Local Within Two Years—No. 2 DBS Provider Said It Will Deliver Local TV Stations to All 210 DMAs During that Time Frame," *Multichannel News*, October 15, 2009.

[32] *CBS Broad. Inc. v. Echostar Comm. Corp.*, 11th Cir. Docket No. 03-13671 (May 23, 2006).

[33] *In the Matter of Application of DISH Network, LLC for Qualified Carrier Certification*, MB Docket No. 10-124, adopted on September 1, 2010, and released on September 2, 2010.

[34] "Media Bureau Seeks Comment for Report Required by the Satellite Television Extension and Localism Act on In-State Broadcast Programming," FCC Public Notice, MB Docket No. 10-238, DA 10-2227, released November 23, 2010.

[35] See, for example, the written testimony of Derek Chang, executive vice president, content strategy and development, DirecTV, Inc., before the House Committee on Energy and Commerce, Subcommittee on Communication Technology, and the Internet, June 16, 2009, at pp. 5-6.

[36] 17 U.S.C. § 119(a)(15)(B).

[37] 17 U.S.C. § 119(a)(15)(D).

[38] 47 U.S.C. § 335(b)(1).

[39] Written Testimony of Bill Acker, Director of Broadcasting and Technology, West Virginia Public Broadcasting, before the Senate Committee on Commerce, Science, and Transportation, Subcommittee on Communications, Technology and the Internet, October 7, 2009, at p. 3.

[40] Kamala Lane, "Dish Sues FCC Over 'PBS HD Mandate,'" *Satellite Week*, July 12, 2010.

[41] "Satellite TV," *Satellite Week*, July 26, 2010.

[42] John Eggerton, "Dish Averts STELA Carriage Mandate," *Multichannel News*, August 2, 2010.

[43] Ibid. at pp. 8-10.

[44] The statutory provisions for satellite explicitly require the use of Nielsen's DMAs. (17 U.S.C. § 122(j)(2)(A) and (C).) The statutory provisions for cable instructed the FCC to make market determinations "using, where available, commercial publications which delineate television markets based on viewing patterns." (47 U.S.C. § 534(h)(1)(C).) Nielsen had already delineated such television markets, assigning geographic areas to markets based on predominant viewing patterns in order to construct ratings data for advertisers, and the FCC therefore adopted Nielsen's market delineations.

[45] For a complete state-by-state list of these counties, their populations, and the full power television stations located in the counties, see the Appendix to CRS Report R40624, *Reauthorizing the Satellite Home Viewing Provisions in the Communications Act and the Copyright Act: Issues for Congress*, by Charles B. Goldfarb.

[46] 17 U.S.C. §§ 119(a)(2)(c)(i)-(iv) and 47 U.S.C. § 341.

[47] See, for example, John Eggerton, "Affiliate Associations Warn Legislators Against Allowing Imported Signals from In-State, Distant Markets," *Broadcasting & Cable,* March 30, 2009.

[48] Representative Ross had introduced the Local Television Freedom Act of 2009, which would have allowed multichannel video programming distributors (MVPDs)—satellite operators and cable operators (including telephone companies)—serving an orphan county to retransmit to their subscribers in that county the signals of television broadcast stations located in an adjacent in-state market. In addition, the Four Corners Television Access Act of 2009 had been introduced in both the House (by Representatives Salazar and Coffman) and the Senate (by Senators Bennet and Udall) to allow satellite operators to retransmit the signals of certain in-state broadcast stations to subscribers located in two Colorado counties that are assigned to the Albuquerque, NM, local market and to allow cable operators located in those counties to retransmit the signals of certain in-state stations without having to obtain retransmission consent from the stations. Also, Representative Boren had introduced a bill which would have allowed satellite operators to retransmit to any subscriber in the state of Oklahoma—not just those in orphan counties—the signals of any broadcast station located in that state.

[49] See Anandashankar Mazundar, "Senate Judiciary Committee Votes Out Satellite Television Reauthorization Bill," *BNA Daily Report for Executives*, September 25, 2009.

[50] "Media Bureau Seeks Comment for Report Required by the Satellite Television Extension and Localism Act on In-State Broadcast Programming," MB Docket No. 10-238, DA 10-2227, released November 23, 2010.

[51] *Satellite Home Viewer Extension and Reauthorization Act Section 109 Report,* A Report of the Register of Copyrights, June 2008, at p. xiv.

[52] See, for example, the Written Testimony of Robert Gabrielli, senior vice president for program operations, DirecTV, Inc., before the Senate Committee on Commerce, Science, and Transportation, October 7, 2009, at p. 8.

[53] See, for example, the website of Public Knowledge at http://www.publicknowledge.org /issues/copyright.

[54] Prior to this clarification, there have been situations in which a cable operator has been required to make a copyright payment as if it were retransmitting a distant signal to all the communities in a service area, but in fact was not allowed to retransmit the signal to certain communities in the service area. Cable operators have referred to the signals that they were not allowed to retransmit, but for which they had to make copyright payments, as "phantom signals."

INDEX

A

B

C

D

E

F

S

T

U

V

W